D0153037

Data Analysis for Chemists

Data Analysis for Chemists

Applications to QSAR and Chemical Product Design

David Livingstone
Independent Consultant and Visiting Professor at the Centre for Molecular Design, University of Portsmouth, UK.
Formerly Manager, QSAR, SmithKline Beecham Pharmaceuticals

Oxford New York Tokyo
OXFORD UNIVERSITY PRESS
1995

Oxford University Press, Walton Street, Oxford OX2 6DP

Oxford New York
Athens Auckland Bangkok Bombay
Calcutta Cape Town Dar es Salaam Delhi
Florence Hong Kong Istanbul Karachi
Kuala Lumpur Madras Madrid Melbourne
Mexico City Nairobi Paris Singapore
Taipei Tokyo Toronto

and associated companies in
Berlin Ibadan

Oxford is a trade mark of Oxford University Press

Published in the United States
by Oxford University Press Inc., New York

A catalogue record for this book is available from the British Library

Library of Congress Cataloging in Publication Data
Livingstone, David (David J.)
Data analysis for chemists: applications to QSAR and chemical product design/David Livingstone
Includes bibliographical references and index.
1. QSAR (Biochemistry)—Statistical methods. I. Title.
QP517.S85L55 1995 574.19'2'072—dc20 95-14967

ISBN 0 19 855728 0

Typeset by Cotswold Typesetting Ltd, Gloucester

Printed in Great Britain by
Bookcraft (Bath) Ltd, Midsomer Norton, Avon

This is for my sister, Delia
(13th July 1942–26th July 1990)

Preface

The inspiration for this book came in part from teaching quantitative aspects of drug design to the B.Sc. and M.Sc. students of medicinal chemistry at the University of Sussex. It has also been necessary for me to describe a number of mathematical and statistical methods to my friends and colleagues in medicinal (and physical) chemistry, biochemistry, and pharmacology departments at Wellcome Research and SmithKline Beecham Pharmaceuticals. I have looked for a textbook which I could recommend which gives *practical* guidance in the use and interpretation of the apparently esoteric methods of multivariate statistics, otherwise known as pattern recognition or chemometrics. I would have found such a book useful when I was learning the trade, and so this is intended to be that sort of guide.

There are, of course, many fine textbooks of statistics, as there are for medicinal chemistry and quantitative drug design, and these are referred to as appropriate for further reading. However, I feel that there isn't a book which gives a practical guide for chemists to the processes of data analysis. The emphasis here is on the application of the techniques, although a certain amount of theory is required in order to explain the methods. This is not intended to be a statistical textbook, indeed an elementary knowledge of statistics is assumed of the reader, but is meant to be a statistical companion to the novice or casual user.

It is necessary here to consider the type of research which these methods may be used for. Historically, techniques for building models, both physical and mathematical, to relate biological properties to chemical structure have been developed in pharmaceutical and agrochemical research. Many of the examples used in this text are derived from these fields of work. There is no reason, however, why any sort of property which depends on chemical structure should not be modelled in this way. This might be termed quantitative structure–property relationships (QSPR) rather than QSAR where A stands for activity. Such models are beginning to be reported; recent examples include applications in the design of dyestuffs, cosmetics, egg-white substitutes, artificial sweeteners, cheese-making, and prepared food products. I have tried to incorporate some of these applications to illustrate the methods, as well as the more traditional examples of QSAR.

The chapters are ordered in a logical sequence, the sequence in which data analysis might be carried out—from planning an experiment through examining and displaying the data to constructing quantitative models. However, each chapter is intended to stand alone so that casual users can refer to the section that is most appropriate to their problem. The one exception to this is the introduction which explains many of the terms which are used later in the book. Finally, I have included definitions and descriptions of some of the chemical properties and biological terms used in panels separated from the rest of the text. Thus, a reader who is already familiar with such concepts should be able to read the book without undue interruption.

Steeple Morden D.J.L.
July 1995

Acknowledgements

I am grateful to the B.Sc. and M.Sc. students, for asking the most difficult questions, and to my research students and colleagues, for asking the most persistent ones. I am very grateful to my friends and colleagues at the University of Portsmouth—Martyn Ford, Richard Greenwood, and David Salt for taking the time and trouble to read this book and make many helpful suggestions. Similarly, I am grateful to John Wood (Wellcome Research), Jan Langowski (LHASA UK), and Kwan Lee (SmithKline Beecham, Philadelphia) for their valuable comments. However, as is customary, I claim the credit for all mistakes and errors.

I would like to thank Sue Brown of Oxford University Press for her very helpful comments on my manuscript and for making the whole process of producing this book much less painful than I feared. Finally, and most importantly, none of this would have been possible without my wife Cherry. For all her hard work, constant encouragement, correction of my poor English and, above all, sacrifice of her precious time—**Thank You.**

Contents

Abbreviations

π	hydrophobicity substituent constant
σ	electronic substituent constant
Λ_{alk}	hydrogen-bonding capability parameter
ΔH	enthalpy
AI	artificial intelligence
ANN	artificial neural networks
ANOVA	analysis of variance
CA	cluster analysis
CAMEO	computer assisted mechanistic evaluation of organic reactions
CASE	computer assisted structure evaluation
CCA	canonical correlation analysis
CONCORD	CONnection table to CoORDinates
CSA	cluster significance analysis
DEREK	deductive estimation of risk from existing knowledge
ED_{50}	dose to give 50 per cent effect
ESDL10	electrophilic superdelocalizability
ESS	explained sum of squares
FA	factor analysis
FOSSIL	frame orientated system for spectroscopic inductive learning
GABA	γ-aminobutyric acid
GC–MS	gas chromatography–mass spectrometry
HPLC	high-performance liquid chromatography
I_{50}	concentration for 50 per cent inhibition
IC_{50}	concentration for 50 per cent inhibition
ID3	iterative dichotomizer three
K_m	Michaelis–Menten constant
KNN	*k*-nearest neighbour technique
LC_{50}	concentration for 50 per cent lethal effect
LD_{50}	dose for 50 per cent death
LDA	linear discriminant analysis
LLM	linear learning machine
log *P*	logarithm of a partition coefficient
LOO	leave one out at a time
LV	latent variable
m.p.	melting point
MAO	monoamine oxidase
MIC	minimum inhibitory concentration
MLR	multiple linear regression
mol.wt.	molecular weight

MR	molar refractivity
MSD	mean squared distance
MSE	explained mean square
MSR	residual mean square
MTC	minimum threshold concentration
NLM	non-linear mapping
NMR	nuclear magnetic resonance
NOA	natural orange aroma
NTP	National Toxicology Program
OLS	ordinary least squares
PC	principal component
PCA	principal component analysis
PCR	principal component regression
p.d.f.	probability density function
pI_{50}	negative log of the concentration for 50 per cent inhibition
PLS	partial least squares
PRESS	predicted residual sum of squares
QDA	quantitative descriptive analysis
QSAR	quantitative structure–activity relationship
QSPR	quantitative structure–property relationship
R^2	multiple squared correlation coefficient
ReNDeR	Reversible Non-linear Dimension Reduction
RSS	residual or unexplained sum of squares
SE	standard error
SAR	structure–activity relationships
SIMCA	see footnote p. 145
SMA	spectral map analysis
SMILES	simplified molecular input line entry system
TD_{50}	dose for 50 per cent toxic effect
TOPKAT	toxicity prediction by komputer assisted technology
TSD	total squared distance
TSS	total sum of squares
UV	ultraviolet spectrophotometry
V_m	Van der Waals volume

1
Chemical properties and chemical structure

1.1 Introduction

Most applications of data analysis involve attempts to fit a model, usually quantitative,* to a set of experimental measurements or observations. The reasons for fitting such models are varied. For example, the model may be purely empirical and be required in order to make predictions for new experiments. On the other hand, the model may be based on some theory or law, and an evaluation of the fit of the data to the model may be used to give insight into the processes underlying the observations made. In some cases the ability to fit a model to a set of data successfully may provide the inspiration to formulate some new hypothesis. The type of model which may be fitted to any set of data depends not only on the nature of the data (see Chapter 3) but also on the intended use of the model. In many applications a model is meant to be used predictively, but the predictions need not necessarily be quantitative. Chapters 4 and 5 give examples of techniques which may be used to make qualitative predictions, as do the classification methods described in Chapter 7.

In some circumstances it may appear that data analysis is not fitting a model at all! The simple procedure of plotting the values of two variables against one another might not seem to be modelling, unless it is already known that the variables are related by some law (for example absorbance and concentration, related by Beer's law). The production of a bivariate plot may be thought of as fitting a model which is simply dictated by the variables. This may be an alien concept but it is a useful way of visualizing what is happening when multivariate techniques are used for the display of data (see Chapter 4). The resulting plots may be thought of as models which have been fitted by the data and as a result they give some insight into the information that the model, and hence the data, contains.

How can such models be constructed to describe the relationship between biological activity (or some other property) and chemical structure?

---

* According to the type of data involved, the model may be qualitative.

1.2 What is QSAR/QSPR?

The fact that different chemicals have different biological effects has been known for millenniums; perhaps one of the earliest examples of a medicine was the use by the ancient Chinese of Ma Haung, which contains ephedrine, to treat asthma and hay fever. Table 1.1 lists some important biologically active materials derived from plants; no doubt most readers will be aware of other bioactive substances derived from plants. Of course it was not until the science of chemistry had become sufficiently developed to assign structures to compounds that it became possible to begin to speculate on the cause of such biological properties. The ability to determine structure enabled early workers to establish structure–activity relationships (SAR), which are simply observations that a certain change in chemical structure has a certain effect on biological activity. As an example, molecules of the general formula shown in Fig. 1.1 are active against the malaria parasite, *Plasmodium falciparum*. The effect of structural changes on the biological properties of derivatives of this compound are shown in Table 1.2, where the chemotherapeutic index is the ratio of maximum tolerated dose to minimum therapeutic dose.

Table 1.1 Some examples of plant-derived compounds

Artemisin	Antimalarial	Sweet wormwood *(Artemisia annua L.)*
Ascaridol	Anthelminthic	Jerusalem artichoke *(Chemopodium anthelminticum)*
Aspirin	Analgesic	Willow bark *(Salix sp.)*
Caffeine	Stimulant	Tea leaves and coffee beans
Digitalis	Antiarrythmic	Foxglove *(Digitalis purpurea)*
Ephedrine	Sympathomimetic	Ma Huang *(Ephedra sinica)*
Filicinic acid	Anthelminthic	Fern *(Aspidium filix mas)*
Nicotine	Stimulant	Tobacco *(Nicotiana tabacum)*
Permethrin	Insecticide	Crysanthemum
Quinine	Antimalarial	Cinchona bark *(Chinchona officinalis)*
Reserpine	Tranquilizer sedative	Fern *(Rauvolfia spp.)*
Strychnine	Central nervous system stimulant	Seeds *(Strychnus nux vomica)*
Taxol	Antitumour	Pacific yew tree *(Taxus breviofolia)*
Vinblastin and Vincristine	Antitumour	Rosy periwinkle *(Catharanthus roseus)*

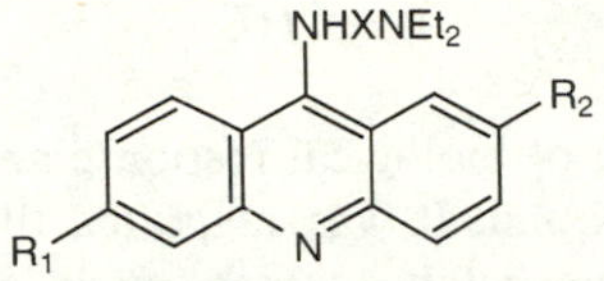

Fig. 1.1. Parent structure of the antimalarial compounds in Table 1.2.

Such relationships are empirical and are semi-quantitative in that the effect of changes in structure are represented as 'all or nothing' effects. In the example shown above, replacement of oxygen by sulphur (compounds 8 and 3) results in a decrease in activity by a factor of 5, but that is all that can be said about that particular chemical change. In this case there is only one example of that particular substitution and thus it is not possible to predict anything other than the fivefold change in activity. If the set of known examples contains a number of such changes then it would be possible to determine a mean effect for this substitution and also to assign a range of likely changes in activity for the purposes of prediction.

Table 1.2 Effect of structural variation on the antimalarial activity of derivatives of the parent compound shown in Fig. 1.1

	X	R1	R2	Chemotherapeutic index
1	$(CH_2)_2$	NO_2	OEt	0
2	$(CH_2)_2$	Cl	OMe	8
3	$(CH_2)_3$	Cl	OMe	15
4	$(CH_2)_3$	H	H	0
5	$(CH_2)_3$	Cl	OEt	7.5
6	$(CH_2)_4$	Cl	OEt	11.2
7	$(CH_2)_3$	CN	OMe	10
8	$(CH_2)_3$	Cl	SMe	2.8

An SAR such as that shown here only applies to the set of compounds from which it is derived, the so-called 'training set' as discussed in Section 1.5 and Chapter 2. Although this might be seen as a disadvantage of structure-activity relationships, the same qualification also applies to other quantitative models of the relationship between structure and activity. One of the powerful features of modelling is also one of its disadvantages, in that any model can only be as 'good' as the training set used to derive it. Making use of a number of more or less reasonable assumptions, the SAR approach has been used to derive more quantitative models of the relationship between structure and activity using a technique known as the Free and Wilson method which is described in Chapter 6.

What then of quantitative structure–activity relationships (QSAR)? The earliest expression of a quantitative relationship between activity and chemical structure was published by Crum Brown and Frazer (1868–9)

$$\phi = f(C) \tag{1.1}$$

where ϕ is an expression of biological response and C is a measure of the 'constitution' of a compound. It was suggested that a chemical operation could be performed on a substance which would produce a known change in its constitution, ΔC. The effect of this change would be to produce a change in its physiological action, $\Delta\phi$. By application of this method to a sufficient number of substances it was hoped that it might be possible to determine what function ϕ is of C. It was recognized that the relationship might not be a strictly mathematical one because the terms ΔC, ϕ, and $\phi + \Delta\phi$ could not be expressed with 'sufficient definiteness to make them the subjects of calculation'. It was expected, however, that it might be possible to obtain an approximate definition of f in eqn (1.1). The key to the difference between the philosophy of this approach and SAR lies in the use of the term quantitative. The Q in QSAR refers to the way in which chemical structures are described, using quantitative physicochemical descriptors. It does not refer to the use of quantitative measures of biological response, although this is a common misconception.

Perhaps the most famous examples of early QSAR are seen in the linear relationships between the narcotic action of organic compounds and their oil/water partition coefficients (Meyer 1899; Overton 1899). Table 1.3 lists the anaesthetic activity of a series of alcohols along with a parameter, $\sum\pi$, which describes their partition properties (see Box 1.2 in this chapter for a description of π). The relationship between this activity and the physicochemical descriptor can be expressed as a linear regression equation as shown below.

$$\log 1/C = 1.039 \sum\pi - 0.442 \tag{1.2}$$

Regression equations and the statistics which may be used to describe their 'goodness of fit', to a linear or other model, are explained in detail in

Table 1.3 Anaesthetic activity and hydrophobicity of a series of alcohols (from Hansch *et al.* 1965, with permission of Academic Press)

Alcohol	$\sum\pi$	Anaesthetic activity ($\log 1/C$)
C_2H_5OH	1.0	0.481
n-C_3H_7OH	1.5	0.959
n-C_4H_9OH	2.0	1.523
n-$C_5H_{11}OH$	2.5	2.152
n-$C_7H_{15}OH$	3.5	3.420
n-$C_8H_{17}OH$	4.0	3.886
n-$C_9H_{19}OH$	4.5	4.602
n-$C_{10}H_{21}OH$	5.0	5.00
n-$C_{11}H_{23}OH$	5.5	5.301
n-$C_{12}H_{25}OH$	6.0	5.124

Chapter 6. For the purposes of demonstrating this relationship it is sufficient to say that the values of the logarithm of a reciprocal concentration (log $1/C$) in eqn (1.2) are obtained by multiplication of the $\sum\pi$ values by a coefficient (1.039) and the addition of a constant term (-0.442). The equation is shown in graphical form (Fig. 1.2); the slope of the fitted line is equal to the regression coefficient (1.039) and the intercept of the line with the zero point of the x-axis is equal to the constant (-0.442).

The origins of modern QSAR may be traced to the work of Professor Corwin Hansch who in the early 1960s proposed that biological 'reactions' could be treated like chemical reactions by the techniques of physical organic chemistry (Hansch *et al.* 1963). Physical organic chemistry, pioneered by Hammett (1937), had already made great progress in the quantitative description of substituent effects on organic reaction rates and equilibria. The best-studied and most well-characterized substituent property was the electronic effect, described by a substituent constant SIGMA (see Box 1.1). Hansch, however, recognized the importance of partition effects in any attempt to describe the properties of compounds in a biological system. The reasoning behind this lay in the recognition that in order to exert an effect on a system, a compound first had to reach its site of action. Since biological systems are composed of a variety of more or less aqueous phases separated by membranes, measurement of partition coefficients in a suitable system of immiscible solvents might provide a simple chemical model of these partition steps in the biosystem.* Although the olive oil/water partition system had already been demonstrated to be of utility, Hansch chose octan-1-ol as the organic phase of his chemical model system of partition. Octan-1-ol was chosen for a variety of

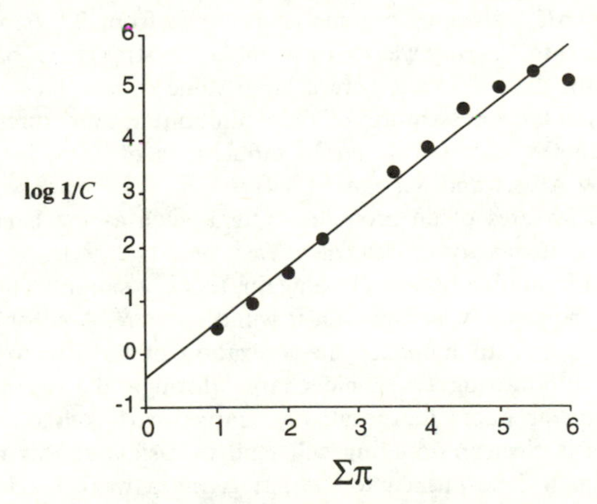

Fig. 1.2. Plot of biological response (log $1/C$) against $\sum\pi$ (from Table 1.3).

* The organic phase of a partition coefficient system is intended to model the fatty, hydrophobic (water hating), membranes and the aqueous phase the hydrophilic parts of a biosystem.

Box 1.1

The electronic substituent constant, σ

Consider the ionization of benzoic acid as shown below where X is a substituent in the *meta* or *para* position to the carboxyl group.

$$\text{X-C}_6\text{H}_4\text{-COOH} \rightleftharpoons \text{X-C}_6\text{H}_4\text{-COO}^{\ominus} + \text{H}^+$$

The extent to which this equilibrium goes to the right, to produce the carboxylate anion and a proton, may be expressed by the value of the equilibrium constant, K_a^c, which is known as the concentration ionization constant

$$K_a^c = \frac{[A^-][H^+]}{[HA]}$$

where the terms in square brackets represent the molar concentrations of the ionized acid (A^-), protons (H^+), and the un-ionized acid (HA). This is a simplification of the treatment of ionization and equilibria but will serve for the purposes of this discussion. The 'strength' of an organic* acid, i.e. the extent to which it ionizes to produce protons, is given by the magnitude of K_a, most often expressed as the negative logarithm of K_a, pK_a. Since pK_a uses the negative log, a large value of K_a will lead to a small number and vice versa. Typical pK_a values of organic acids range from 0.5 (strong) for trifluoroacetic acid to 10 (very weak) for phenol. The strength of bases can also be expressed on the pK_a scale; here a large value of pK_a indicates a strong base. A very readable description of the definition and measurement of acid and base strengths, along with useful tabulations of data, is given in the monograph by Albert and Serjeant (1984).

One of the features of an aromatic system, such as the benzene ring in benzoic acid, is its ability to delocalize electronic charge through the alternating single and double bonds. Once again, this is a simplification, since the bonds are all the same type; however, it will serve here. A substituent on the benzene ring is able to influence the ionization of the carboxyl group by donating or withdrawing electronic charge through the aromatic system. Since ionization produces the negatively charged carboxylate anion, a substituent which is electron-donating will tend to disfavour this reaction and the equilibrium will be pushed to the left giving a weaker acid, compared with the unsubstituted, with a higher pK_a. An electron-withdrawing sub-

---

* Inorganic acids, such as HCl, H_2SO_4, and HNO_3, are effectively always completely dissociated in aqueous solution.

stituent, on the other hand, will tend to stabilize the anion since it will tend to 'spread' the negative charge and the equilibrium will be pushed to the right resulting in a stronger acid than the unsubstituted parent. Hammett (1937) reasoned that the effect of a substituent on a reaction could be characterized by a substituent constant, for which he chose the symbol σ and a reaction constant, ρ. Thus, for the ionization of benzoic acids the Hammett equation is written as

$$\rho\sigma_x = \log K_x - \log K_H$$

where the subscripts x and H refer to an x substituent and hydrogen (the parent) respectively. Measurement of the pK_a values of a series of substituted benzoic acids and comparison with the parent leads to a set of $\rho\sigma$ products. Choice of a value of ρ for a given reaction allows the extraction of σ values; Hammett chose the ionization of benzoic acids at 25°C in aqueous solution as a standard since there was a large quantity of accurate data available. This reaction was given a ρ value of 1; the substituent σ values derived from these pK_a measurements have been successfully applied to the quantitative description of many other chemical equilibria and reactions.

reasons: perhaps the most important is that it consists of a long hydrocarbon chain with a relatively polar hydroxyl head group, and therefore mimics some of the lipid constituents of biological membranes. The octanol/water system has provided one of the most successful physicochemical descriptors used in QSAR, although arguments have been proposed in favour of other models and, recently, three further chemical models of partition have been suggested (Leahy *et al.* 1989). It was proposed that these provide information that is complementary to that of the octanol/water system. When Hansch first published on the octanol/water system he defined (Hansch *et al.* 1962) a substituent constant, π, in an analogous fashion to the Hammett σ constant (see Box 1.2). The generalized form of what has now become known as the Hansch approach is shown below

$$\log 1/C = a\pi + b\pi^2 + c\sigma + dE_s + \text{const.} \quad (1.3)$$

where C is the dose required to produce a standard effect (see Section 1.4.1); π, σ, and E_s are hydrophobic, electronic, and steric parameters respectively (see Box 1.3); a, b, c, and d are coefficients fitted by regression; and const. is a constant. The squared term in π is included in an attempt to account for non-linear relationships in hydrophobicity. The form of an equation with a squared term is a parabola and it is true that a number of data sets *appear* to fit a parabolic relationship with the partition coefficient. However, a number of other non-linear relationships may also be fitted to such data sets and non-linear modelling in hydrophobicity has received some attention, as described in Chapter 6.

Box 1.2

The hydrophobic substituent constant, π

The partition coefficient, P, is defined as the ratio of the concentrations of a compound in the two immiscible phases used in the partitioning system. The custom here is to take the concentration in the organic phase as the numerator; for most QSAR applications the organic phase is 1-octanol.

$$P = \frac{[\ ]_{OCT}}{[\ ]_{AQ}}$$

Here, the terms in the square brackets refer to the concentration of the same species in the two different phases.

Hansch chose logarithms of the partition coefficients of a series of substituted benzenes to define a substituent constant, π, thus

$$\pi_x = \log P_x - \log P_H$$

where x and H refer to an x-substituted benzene and the parent, benzene, respectively. The similarity with the Hammett equation may be seen but it should be noted that there is no reaction constant equivalent to the Hammett constant ρ. If a substituent has no effect on the partitioning properties of benzene, its π value will be zero. If it increases partition into the octanol phase, then P, and hence log P, will be larger than for benzene and π will be positive. Such a substituent is said to be hydrophobic; a substituent which favours partition into the aqueous phase will have a negative π value and is said to be hydrophilic. Some representative π values are shown in the table.

Hydrophobic		Hydrophilic	
Substituent	π	Substituent	π
$-CH_3$	0.56	$-NO_2$	−0.28
$-C(CH_3)_3$	1.98	$-OH$	−0.67
$-C_6H_5$	1.96	$-CO_2H$	−0.32
$-C_6H_{11}$	2.51	$-NH_2$	−1.23
$-CF_3$	0.88	$-CHO$	−0.65

A couple of interesting facts emerged from early investigations of π values following the measurement of partition coefficients for several series of compounds. The substituent constant values were shown to be more or less constant and their effects to be, broadly speaking, additive. This is of particular importance if a quantitative relationship involving π is to be used predictively; in order to predict activity it is necessary to be able to predict values for the substituent parameters. Additivity breaks down when there are interactions between substituents, e.g. steric interactions, hydrogen bonds, electronic effects; 'constancy' breaks down when there is some interaction (usually electronic) between the substituent and the parent structure. One way to avoid any such problem is to use whole molecule log P values, or log P for a fragment of interest, but of course this raises the question of calculation for the purposes of prediction. Fortunately, log P values may also be calculated and there are a number of more or less empirical schemes available for this purpose (Livingstone 1991).

In the defining equation for the partition coefficient it was noted that the concentration terms referred to the concentration of the same species. This can have significance for the measurement of log P if some interaction (for example, dimerization) occurs predominantly in one phase, but is probably of most significance if the molecule contains an ionizable group. Since P refers to one species it is necessary to suppress ionization by the use of a suitable pH for the aqueous phase. An alternative is to measure a distribution coefficient, D, which involves the concentrations of both ionized and un-ionized species, and apply a correction factor based on the pK_a values of the group(s) involved. Yet another alternative is to use log D values themselves as a hydrophobic descriptor, although this may suffer from the disadvantage that it includes electronic information.

The measurement, calculation, and interpretation of hydrophobic parameters has been the subject of much debate. For further reading see Leo *et al.* (1971), Dearden and Bresnen (1988), and Livingstone (1991).

Box 1.3

The bulk substituent constant, MR

In eqn (1.3) three substituent parameters π, σ, and E_s, are used to describe the hydrophobic, electronic, and steric properties of substituents. The substituent constant E_s, due to Taft (1956), is based on the measurement of rate constants for the acid hydrolysis of esters of the following type

$$X\text{–}CH_2COOR$$

and it is assumed that the size of the substituent X will affect the ease with which a transition state in the hydrolysis reaction is achieved.

A variety of successful correlations have been reported in which E_s has been involved but doubt has been expressed as to its suitability as a steric descriptor, mainly due to concern that electronic effects of substituents may predominantly control the rates of hydrolysis. Another problem with the E_s parameter is that many common substituents are unstable under the conditions of acid hydrolysis.

An alternative parameter for 'size', molar refractivity (MR), was suggested by Pauling and Pressman (1945). MR is described by the Lorentz–Lorenz equation

$$MR = \frac{n^2 - 1}{n^2 + 1} \cdot \frac{\text{mol.wt}}{d}$$

where n is the refractive index, and d is the density of a compound, normally a liquid. MR is an additive-constitutive property and thus can be calculated by the addition of fragment values, from look-up tables, and nowadays by computer programs (Livingstone 1991). This descriptor has been successfully employed in many QSAR reports although, as for E_s, debate continues as to precisely what chemical property it models. A variety of other parameters has been proposed for the description of steric/bulk effects (Livingstone 1991) including various corrected atomic radii (Charton 1991).

What of QSPR? In principle, any property of a substance which is dependent on the chemical properties of one or more of its constituents could be modelled using the techniques of QSAR. Although most of the reported applications come from pharmaceutical and agrochemical research, publications from more diverse fields are beginning to appear. For example, Narvaez and co-workers (1986) analysed the relationship between musk odourant properties and chemical structure for a set of bicyclo- and tricyclobenzenoids. A total of 47 chemical descriptors were generated (Table 1.4) for a training set of 148 compounds comprising 67 musks and 81 non-musks. Using the final set of 14 parameters, a discriminant function (see Chapter 7) was generated which was able to classify correctly all of the training set compounds. A test set of 15 compounds, six musk and nine non-musks, was used to check the predictive ability of the discriminant functions. This gave correct predictions for all of the musk compounds and eight of the nine non-musks.

Another example involves a quantitative description of the colourfastness of azo dye analogues of the parent structure shown in Fig. 1.3

Table 1.4 Descriptors used in the analysis of musks (from Narvaez *et al.* 1986, with permission of Oxford University Press)

	Number of descriptors		
	Generated	Used[a]	Final[b]
Substructure	8	7	2
Substructure environment	6	6	2
Molecular connectivity	9	6	4
Geometric	15	8	4
Calculated log P	1	1	0
Molar refractivity	1	1	0
Electronic	7	6	2
Total	47	35	14

[a] Some descriptors were removed prior to the analysis due to correlations with other parameters or insufficient non-zero values (See Chapter 3).
[b] Number of parameters used in the final predictive equation.

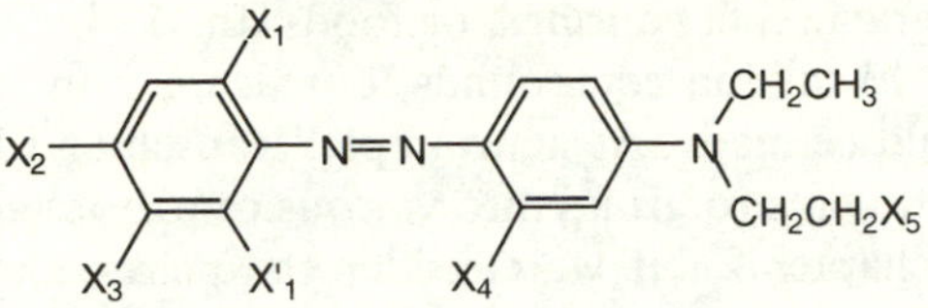

Fig. 1.3. Parent structure of azo dye analogues (from Carpignano *et al.* 1985 with permission of the Society of Dyers and Colourists).

(Carpignano *et al.* 1985). Amongst other techniques, this study applied the method of Free and Wilson (see Chapter 6) to the prediction of colour-fastness. Briefly, the Free and Wilson method involves the calculation, using regression analysis, of the contribution that a substituent in a particular position makes to activity, here the activity is light-fastness of the dye. It is assumed that substituents make a constant contribution to the property of interest and that these contributions are additive. The analysis gave a regression equation which explained 92 per cent of the variance in the light-fastness data with a standard deviation of 0.49. An extract of some of the predictions made by the Free and Wilson analysis is shown in Table 1.5, which includes the best and worse predictions and also shows the range of the data. One advantage of this sort of treatment of the data is that it allows the identification of the most important positions of substitution (X_1 and X_5) and the most positively (CN and Cl) and negatively influential substituents (NO_2 and OCH_3).

Table 1.5 Predicted light-fastness of azo dyes (from Carpignano *et al.* 1985, with permission of the Society of Dyers and Colourists)

Dye[a]	Calculated light-fastness	Residual[b]
1	4.03	0.47
4	5.50	0.00
8	5.35	−0.35
13	2.84	0.16
18	1.05	−0.05
19	6.69	0.31
21	4.25	0.75
28	2.15	−0.15
39	5.76	−0.76
44	5.36	−0.36

[a] Selected dyes from a larger set have been shown here.
[b] Difference between predicted and measured.

1.3 Why look for quantitative relationships?

The potential of organic chemistry for the production of new compounds is enormous, whether they be intended for pharmaceutical or agrochemical applications, fragrances, flavourings, or foods. In 1994, *Chemical Abstracts* listed more than 13 million compounds, but this is only a tiny percentage of those that could be made. As an example, Hansch and Leo (1979) chose a set of 166 substituents to group into various categories according to their properties (see Chapter 2). If we consider the possible substitution positions on the carbon atoms of a relatively simple compound such as quinoline (Fig. 1.4), there are 10^{15} different analogues that can be made using these substituents. If the hunt for new products merely involved the synthesis and testing of new compounds without any other guidance, then it would clearly be a long and expensive task.

Of course, this is not the way that industry goes about the job. A large body of knowledge exists ranging from empirical structure–activity relationships to a detailed knowledge of mechanism, including metabolism and elimination in some cases. The purpose of quantitative structure–activity (or property) relationships is to provide a better description of chemical structure and perhaps some information concerning mechanism. The advantage of having a better description of structure is that it may be possible to transfer information from one series to another. In the example shown in Section 1.2, it was seen that substitution of a sulphur atom by oxygen resulted in an improvement in activity. This may be due to a change in lipophilicity, bulk, or electronic properties. If we know which parameters are important then we can, within the constraints of organic chemistry, design molecules which have the desired properties by making changes which are more significant than swapping oxygen for sulphur.

The work of Hansch *et al.* (1977) provides an example of the use of QSAR to give information concerning mechanism. They demonstrated the following relationship for a set of esters binding to the enzyme papain.

$$\log 1/K_m = 1.03\pi'_3 + 0.57\sigma + 0.61\mathrm{MR}_4 + 3.8 \qquad (1.4)$$
$$n = 25 \quad r = 0.907 \quad s = 0.208$$

Where K_m, the Michaelis–Menten constant, is the substrate concentration at which the velocity of the reaction is half maximal. The subscripts to the physicochemical parameters indicate substituent positions. The statistics quoted are the number of compounds in the data set (n), the correlation coefficient (r) which is a measure of goodness of fit, and the standard error of the fit (s); see Chapter 6 for an explanation of these statistics. It is possible to try to assign some chemical 'meaning' to the physicochemical

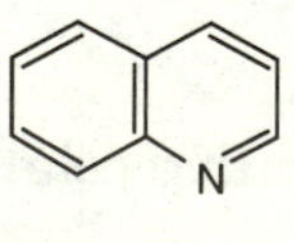

$166^7 = 3.47 \times 10^{15}$

Fig. 1.4. Quinoline

parameters involved in eqn (1.4). The positive coefficient for σ implies that electron-withdrawing substituents favour formation of the enzyme–substrate complex. Since the mechanism of action of papain involves the electron-rich SH group of a cysteine residue, this appears to be consistent. The molar refractivity term (see Box 1.3) is also positive, implying that bulkier substituents in the 4 position favour binding. The two parameters π_4 and MR_4 are reasonably orthogonal* for the set of 25 compounds used to generate eqn (1.4), and since the data does not correlate with π_4 it was concluded that a bulk effect rather than a hydrophobic effect was important at position 4. The prime sign associated with the π parameter for position 3 indicates that where there were two *meta* substituents the π value of the more hydrophobic substituent was used, the other π_3 value being ignored. The rationale for this procedure was that binding of one *meta* substituent to the enzyme placed the other *meta* substituent into an aqueous region outside the enzyme binding site. It was also necessary to make this assumption in order to generate a reasonable regression equation which described the data.

Following the QSAR analysis, Hansch and Blaney (1984) constructed a model of the enzyme and demonstrated that the invariant hydrophobic portion of the molecules could bind to a large hydrophobic pocket. In this model, one of the two *meta* substituents also fell into a hydrophobic pocket forcing the other *meta* substituent out of the binding site. The substituent at the 4 position points towards an amide group on the enzyme which is consistent with the assignment of a bulk not hydrophobic component to enzyme binding at this position. The QSAR equation and molecular graphics study in this instance appear to tie together very nicely and it is tempting to expect (or hope!) that this will always be the case. A note of caution should be sounded here in that strictly speaking a correlation does not imply causality. However, there is no need to be unduly pessimistic: correlation can inspire imagination!

1.4 Sources of data

At this point it is necessary to introduce some jargon which will help to distinguish the two main types of data which are involved in QSAR. The biological or other property data is known as a **dependent variable**. It is expected that this type of data will be determined by chemical structure, and it will thus be related by some more or less complex function to the physicochemical properties which are themselves dictated by structure. It is the aim of QSAR to predict values of one or more dependent variables from values of one or more **independent variables**. The independent

* Orthogonal here means uncorrelated.

variables are physicochemical properties such as π and σ which, although dependent on chemical structure, are not dependent on the biological data. Dependent variables are usually determined by experimental measurement or observation on some (hopefully) relevant test system. This may be a biological system such as a purified enzyme, cell culture, piece of tissue, or whole animal; alternatively it may be a panel of tasters, a measurement of viscosity, or the quantification of colour. Independent variables may be determined experimentally or may be calculated. Important features of data such as scales of measurement, distribution, and scaling are described in Chapter 3. Here we shall just consider the sources of data.

1.4.1 Dependent data

Important considerations for dependent data are that their measurement should be well defined experimentally, and that they should be consistent amongst the compounds (mixtures, samples, products) in a set. This may seem obvious, and of course it is good scientific practice to ensure that an experiment is well controlled, but it is not always obvious that data is consistent, particularly when analysed by someone who did not generate it. Consider the set of curves shown in Fig. 1.5 where biological effect is plotted against concentration. Compounds 1–3 can be seen to be 'well behaved' in that their dose–response curves are of very similar shape and are just shifted along the concentration axis depending on their potency. Curves of this sigmoidal shape are quite typical; common practice is to take 50 per cent as the measure of effect and read off the concentration to achieve this from the dose axis. The advantage of this is that the curve is linear in this region; thus if the ED_{50} (the dose to give 50 per cent effect) has been bracketed by experimental measurements, it simply requires

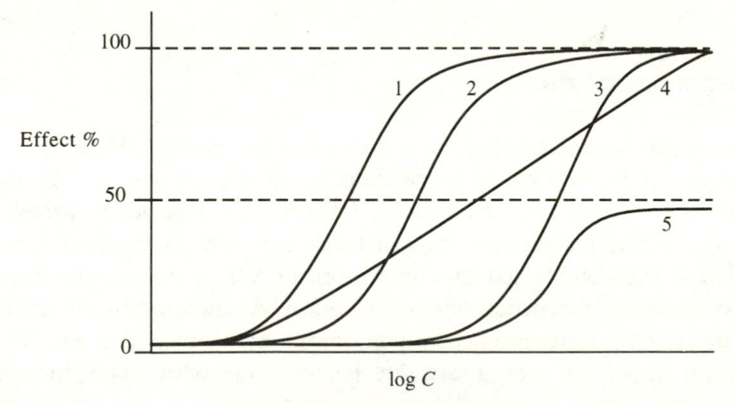

Fig. 1.5. Typical and not so typical dose–response curves for a set of five different compounds.

linear interpolation to obtain the ED_{50}. A further advantage of this procedure is that the effect is changing most rapidly with concentration in the 50 per cent part of the curve. Since small changes in concentration produce large changes in effect it is possible to get the most precise measure of the concentration required to cause a standard effect. The curve for compound 4 illustrates a common problem in that it does not run parallel to the others; this compound produces small effects (<50 per cent) at very low doses but needs comparatively high concentrations to achieve effects in excess of 50 per cent. Compound 5 demonstrates yet another deviation from the norm in that it does not achieve 50 per cent effect. There may be a variety of reasons for these deviations from the usual behaviour, such as changes in mechanism, solubility problems, and so on, but the effect is to produce inconsistent results which may be difficult or impossible to analyse.

The situation shown here where full dose–response data is available is very good from the point of view of the analyst, since it is relatively easy to detect abnormal behaviour and the data will have good precision. However, it is often time-consuming, expensive, or both, to collect such a full set of data. There is also the question of what is required from the test in terms of the eventual application. There is little point, for example, in making precise measurements in the millimolar range when the target activity must be of the order of micromolar or nanomolar. Thus, it should be borne in mind that the data available for a QSAR/QSPR analysis may not always be as good as it appears at first sight. Any time spent in a preliminary examination of the data and discussion with those involved in the measurement will usually be amply repaid.

1.4.2 Independent data

The physicochemical parameters used to describe molecules may come from a variety of sources. Chemical model systems such as those described for hydrophobic and electronic effects have proved successful, as have models for bulk and steric effects. A good account of many of these (mainly) substituent constants is given in the paper by Van de Waterbeemd *et al.* (1989). In principle, any experimentally measured property may be used as a physicochemical descriptor. A number of these are shown in Table 1.6. There can, however, be a major disadvantage in the use of experimental properties. If the experimental parameter cannot be predicted, it will be necessary to synthesize a new compound before the property can be measured. Since a major aim of QSAR is to predict *before* synthesis, the use of such a descriptor may, to some extent, defeat its own purpose. Of course, if the biological test is difficult, slow, or expensive, a QSAR involving experimental parameters may be very useful.

Table 1.6 Some experimental parameters used in QSAR

Descriptor	Symbol
Melting point	m.p.
NMR chemical shift	δ
Infra-red frequency	υ
Half-wave reduction potential	$E_{1/2}$
Half-life for hydrolysis in mouse plasma	$t_{1/2}$
Ionization potential	I_p
Half life for reaction with 4-nitrothiophenol	$t_{1/2}$
Chromatographic retention	R_m

Some experimental properties can be predicted with varying degrees of success. Carbon-13 NMR chemical shifts, for example, may be predicted by use of a combination of modelling by molecular mechanics and consideration of the electronegativity of the constituent atoms of the molecule (Jaime 1990). A handbook of chemical property estimation methods, unfortunately no longer in print, was assembled by Lyman *et al.* (1982).

A very useful addition to these experimental and chemical model-based descriptors are parameters which may be calculated solely from consideration of molecular structure. One such class of parameters is based on the topology of a molecule: they are known as molecular connectivity indices (see Box 1.4). These descriptors may be calculated very readily and have the advantage that they may be applied to quite diverse sets of structures. As a result they have found applications in environmental studies where data sets containing non-homologous structures are not unusual (Basak and Magnuson 1983; De Flora *et al.* 1985; Nirmalakhandan and Speece 1988).

One of the drawbacks with molecular connectivity descriptors is the difficulty in moving back from connectivity values to chemical structure. That is to say, it is difficult to predict a chemical structure which will give a particular value of a connectivity index.

Another class of descriptors which can be calculated purely from a knowledge of molecular structure is obtained from the computer-based

Box 1.4

Molecular connectivity indices

Molecular connectivity is a topological descriptor, that is to say it is calculated from a two-dimensional representation of chemical structure. All that is required in order to calculate molecular connectivity indices for a compound is knowledge of the nature of its constituent atoms (usually just the heavy atoms, not hydrogens) and the way that they are joined to one another.

Consider the hydrogen-suppressed graph of the alcohol shown below. The numbers in brackets give the degree of connectivity, δ_i, for each atom; this is just the number of other atoms connected to an atom. For each bond in the structure, a bond connectivity, C_k, can be calculated by taking the reciprocal of the square root of the product of the connectivities of the atoms at either end of the bond. For example, the bond connectivity for the first carbon–carbon bond (from the left) in the structure is

$$C_1 = 1/\sqrt{1 \times 3}$$

More generally the bond connectivity of the kth bond is given by

$$C_k = 1/\sqrt{\delta_i \delta_j}$$

where the subscripts i and j refer to the atoms at either end of the bond. The molecular connectivity index, χ, for a molecule is found by summation of the bond connectivities over all of its N bonds.

$$\chi = \sum_{k=1}^{N} C_k$$

For the butanol shown below, the four bond connectivities are the reciprocal square roots of (1×3), (1×3), (2×3), and (2×1) which gives a molecular connectivity value of 2.269. This simple connectivity index is known as the first-order index because it considers only individual bonds, in other words paths of two atoms in the structure. Higher order indices may be generated by the consideration of longer paths in a molecule and other refinements

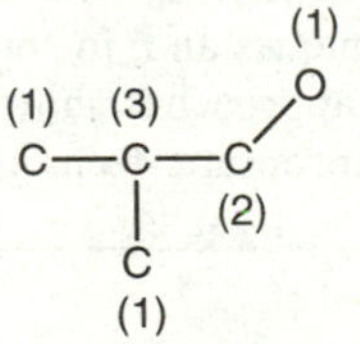

(1)
O
(1) (3)
C—C—C
(2)
C
(1)

have been considered, such as valence connectivity values, path, cluster, and chain connectivities (Kier and Hall 1986).

Molecular connectivity indices have the advantage that they can be readily and rapidly calculated from a minimal description of chemical structure. As might be expected from their method of calculation they contain primarily steric information, although it is claimed that certain indices, particularly valence connectivities, also contain electronic information. Molecular connectivity has been shown to correlate with chemical properties such as water solubility, boiling point, partition coefficient, and Van der Waals' volume. They have also been used to describe a variety of biological properties including toxicity, and have a number of environmental applications.

Box 1.5

Parameters from molecular models

The creation of molecular models using computers has become a common feature in many areas of chemical research. The information used to create the models may be experimental (from X-ray structure determination, NMR experiments, and so on) or may be theoretical (from empirical methods such as molecular mechanics or less empirical techniques such as quantum mechanics). Having built the models, the molecular modelling software may be used to calculate a very wide variety of descriptors. Some may be geometrical, for example, the distance (and spatial relationship) between functional groups, minimum and maximum dimensions in a given axis, volume, etc. Others may depend upon calculation of electronic distribution and energies: molecular orbital energies are good examples, and the highest occupied (HOMO) and lowest unoccupied (LUMO) are particular favourites. Another class of descriptor involves the calculation of an interaction energy between a probe atom or group and the molecule of interest. This is usually carried out at regularly spaced intervals on a grid of points surrounding the molecule or, in the case of a large molecule such as a protein, enclosing a particular part of the computer model. This approach tends to lead to the generation of very large numbers (thousands) of parameters and thus requires multivariate techniques in order to analyse the data. Theoretical descriptors have also been proposed as replacements for 'traditional' parameters such as log P and a grid-based approach has been used to 'explain' pK_a values.

The range of parameters that may be calculated from computer models of chemical structure is only limited by the imagination of the theoretical chemists who devise the techniques and, in some cases, the users who employ them. The success of these approaches, however, is largely determined by correct application of the appropriate techniques of multivariate analysis.

molecular modelling packages which are now widely available. A variety of physicochemical properties may be calculated using such systems (see Box 1.5); Livingstone (1991) discusses the generation of such parameters and also puts them into context with the more 'traditional' descriptors used in QSAR.

1.5 Analytical methods

This whole book is concerned with analytical methods, as the following chapters will show, so the purpose of this section is to introduce and explain some of the terms which are used to describe the techniques. These terms, like most jargon, also often serve to obscure the methodology to the casual or novice user so it is hoped that this section will help to unveil the techniques.

First, we should consider some of the expressions which are used to describe the methods in general. **Chemometrics** is used to describe 'any mathematical or statistical procedure which is used to analyse chemical data' (Kowalski *et al.* 1987). Thus, the simple act of plotting a calibration curve is chemometrics, as is the process of fitting a line to that plot by the method of least squares, as is the analysis by principal components of the spectrum of a solution containing several species. Any chemist who carries out quantitative experiments is also a chemometrician! This term has its analogue in other sciences; biometrics in biology and psychometrics in psychology. **Univariate statistics** is (perhaps unsurprisingly) the term given to describe the statistical analysis of a single variable. This is the type of statistics which is normally taught on an introductory course; it involves the analysis of variance of a single variable to give quantities such as the mean and standard deviation, and some measures of the distribution of the data (see Chapter 3). **Multivariate statistics** describes the application of statistical methods to more than one variable at a time, and is perhaps more useful than univariate methods since most problems in real life are multivariate. We might more correctly use the term **multivariate analysis** since not all multivariate methods are statistical. Chemometrics and multivariate analysis refer to more or less the same things, chemometrics being the broader term since it includes univariate techniques.*

Pattern recognition is the name given to any method which helps to reveal the patterns within a data set. Some of the display techniques described in Chapter 4 are quite obvious examples of pattern recognition since they result in a visual display of the patterns in data. However, to go back to the example of a calibration curve, the equation used to describe the fit of a linear (or otherwise) curve to a data set is also an example of showing a pattern in data. In this case the pattern is not shown visually but is revealed in the rather more concise form of an equation. Pattern recognition and chemometrics are more or less synonymous. Some of the pattern recognition techniques are derived from research into artificial intelligence. We can 'borrow' some useful jargon from this field which is related to the concept of 'training' an algorithm or device to carry out a particular task. Suppose that we have a set of data which describes a collection of compounds which can be classified as active or inactive in some biological test. The descriptor data, or independent variables, may be whole molecule parameters such as melting point, or may be substituent constants such as π, or may be calculated quantities such as molecular

---

* But, of course, it is restricted to chemical problems.

orbital energies. One simple way in which this data may be analysed is to compare the values of the variables for the active compounds with those of the inactives (see discriminant analysis in Chapter 7). This may enable one to establish a rule or rules which will distinguish the two classes. For example, all the actives may have melting points above 250°C and/or may have π values below 1. The production of these rules, by inspection of the data or by use of an algorithm, is called **supervised learning** since knowledge of class membership was used to generate them. The dependent variable, in this case membership of the active or inactive class, is used in the learning or training process. **Unsupervised learning**, on the other hand, does not make use of a dependent variable. An example of unsupervised learning for this data set might be to plot the values of two of the descriptor variables against one another. Class membership for the compounds could then be marked on the plot and a pattern may be seen to emerge from the data. If we chose melting point and π as the two variables to plot, we may see a grouping of the active compounds where $\pi < 1$ and melting point > 250°C.

The distinction between supervised and unsupervised learning may seem unimportant but there is a significant philosophical difference between the two. When we seek a rule to classify data, there is a possibility that any apparent rule may happen by chance. It may, for example, be a coincidence that all the active compounds have high melting points; in such a case the rule will not be predictive. This may be misleading, embarrassing, expensive, or all three! Chance effects may also occur with unsupervised learning but are much less likely since unsupervised learning does not seek to generate rules. Chance effects are discussed in more detail in Chapters 6 and 7. The concept of learning may also be used to define some data sets. A set of compounds which have already been tested in some biological system, or which are about to be tested, is known as a **learning** or **training set**. In the case of a supervised learning method this data will be used to train the technique but this term applies equally well to the unsupervised case. Judicious choice of the training set will have profound effects on the success of the application of any analytical method, supervised or unsupervised, since the information contained in this set dictates the information that can be extracted (see Chapter 2). A set of untested or yet to be synthesized compounds is called a **test set**, the objective of data analysis usually being to make predictions for the test set (also sometimes called a **prediction set**). A further type of data set, known as an **evaluation set**, may also be used. This consists of a set of compounds for which test results are available but which is not used in the construction of the model. Examination of the prediction results for an evaluation set can give some insight into the validity and accuracy of the model.

Finally we should define the terms **parametric** and **non-parametric**. A measure of the distribution of a variable (see Chapter 3) is a measure of one of the parameters of that variable. If we had measurements for all possible values of a variable, (an infinite number of measurements), then we would be able to compute a value for the population distribution. Statistics is concerned with a much smaller set of measurements which forms a sample of that population and for which we can calculate a sample distribution. A well-known example of this is the Gaussian or normal distribution. One of the assumptions made in statistics is that a sample distribution, which we can measure, will behave like a population distribution which we cannot. Although population distributions cannot be measured, some of their properties can be predicted by theory. Many statistical methods are based on the properties of population distributions, particularly the normal distribution. These are called **parametric techniques** since they make use of the distribution parameter. Before using a parametric method, the distribution of the variables involved should be calculated. This is very often ignored, although fortunately many of the techniques based on assumptions about the normal distribution are quite robust to departures from normality. There are also techniques which do not rely on the properties of a distribution, and these are known as **non-parametric** or '**distribution free**' methods.

References

Albert, A. and Serjeant, E. P. (1984). *The determination of ionization constants: a laboratory manual*, (3rd edn). Chapman and Hall, London.

Basak, S. C. and Magnuson, V. R. (1983). *Arzneim–Forsch.*, **33**, 501–3.

Carpignano, R., Savarino, P., Barni, E., Di Modica, G., and Papa, S. S. (1985). *Journal of the Society of Dye Colour*, **101**, 270–6.

Charton, M. (1991). The quantitative description of steric effects. In *Similarity models in organic chemistry, biochemistry and related fields*, (ed. R. I. Zalewski, T.M. Krygowski, and J. Shorter), pp. 629–87. Elsevier, Amsterdam.

Crum Brown, A. and Frazer, T. (1868–9). *Transactions of the Royal Society of Edinburgh*, **25**, 151–203.

Dearden, J. C. and Bresnen, G. M. (1988). *Quantitative Structure–Activity Relationships*, **7**, 133–44.

De Flora, S., Koch, R., Strobel, K., and Nagel, M. (1985). *Toxicological and Environmental Chemistry*, **10**, 157–70.

Hammett, L. P. (1937). *Journal of the American Chemical Society*, **59**, 96–103.

Hansch, C. and Blaney, J. M. (1984). In *Drug Design: Fact or Fantasy?* (ed. G. Jolles and K. R. H. Wooldridge) pp. 185-208. Academic Press, London.

Hansch, C. and Leo, A. (1979). *Substituent constants for correlation analysis in chemistry and biology*. Wiley, New York.

Hansch, C., Maloney, P. P., Fujita, T., and Muir, R. M. (1962). *Nature*, **194**, 178–80.

Hansch, C., Muir, R. M., Fujita, T., Maloney, P. P., Geiger, F., and Streich, M. (1963). *Journal of the American Chemical Society*, **85**, 2817–24.

Hansch, C., Steward, A. R., Iwasa, J., and Deutsch, E. W. (1965). *Molecular Pharmacology*, **1**, 205–13.

Hansch, C., Smith, R. N., Rockoff, A., Calef, D. F., Jow, P. Y. C., and Fukunaga, J. Y. (1977). *Archives of Biochemistry and Biophysics*, **183**, 383–92.

Jaime, C. (1990). *Magnetic Resonance in Chemistry*, **28**, 42–6.

Kier, L. B. and Hall, L. H. (1986). *Molecular connectivity in structure–activity analysis*. Wiley, New York.

Kowalski, B., Brown, S., and Van de Ginste, B. (1987). *Journal of Chemometrics*, **1**, 1–2.

Leahy, D. E., Taylor, P. J., and Wait, A. R. (1989). *Quantitative Structure–Activity Relationships*, **8**, 17–31.

Leo, A., Hansch, C., and Elkins, D. (1971). *Chemical Reviews*, **71**, 525–616.

Livingstone, D. J. (1991). Quantitative structure–activity relationships. In *Similarity models in organic chemistry, biochemistry and related fields*, (ed. R. I. Zalewski, T. M. Krygowski, and J. Shorter), pp. 557–627. Elsevier, Amsterdam.

Lyman, W., Reehl, W., and Rosenblatt, D. (1982). *Handbook of chemical property estimation methods*. McGraw Hill, New York.

Meyer, H. (1899). *Archives of Experimental Pathology and Pharmakology*, **42**, 109.

Meyer, K. H. and Hemmi, H. (1935). *Biochem Z.*, **277**, 39.

Narvaez, J. N., Lavine, B. K., and Jurs, P. C. (1986). *Chemical Senses*, **11**, 145–56.

Nirmalakhandan, N. and Speece, R. E. (1988). *Environmental Science and Technology*, **22**, 606–15.

Overton, E. (1899). *Vierteljahrsschr. Naturforsch. Ges. Zurich*, **44**, 88.

Pauling, L. and Pressman, D. (1945). *Journal of the American Chemical Society*, **67**, 1003.

Taft, R. W. (1956). In *Steric effects in organic chemistry*, (ed. M. S. Newman), p.556, Wiley, New York.

Van de Waterbeemd, H., El Tayar, N., Carrupt, P. A., and Testa, B (1989). *Journal of Computer-Aided Molecular Design*, **3**, 111–32.

2
Experimental design—compound and parameter selection

2.1 What is experimental design?

All experiments are designed insofar as decisions are made concerning the choice of apparatus, reagents, animals, analytical instruments, temperature, solvent, and so on. Such decisions need to be made for any individual experiment, or series of experiments, and will be based on prior experience, reference to the literature, or perhaps the whim of an individual experimentalist. How can we be sure that we have made the right decisions? Does it matter whether we have made the right decisions? After all, it can be argued that an experiment is just that; the results obtained with a particular experimental set-up are the results obtained, and as such are of more or less interest depending on what they are. To some extent the reason for conducting the experiment in the first place may decide whether the question of the right decisions matters. If the experiment is being carried out to comply with some legislation from a regulatory body (for example, toxicity testing for a new drug may require administration at doses which are fixed multiples of the therapeutic dose), then the experimental decisions do not matter. Alternatively the experiment may be intended to synthesize a new compound. In this case, if the target compound is produced then all is well, except that we do not know that the yield obtained is the best we could get by that route. This may not matter if we are just interested in having a sample of the compound, but what should we do if the experiment does not produce the compound? The experiment can be repeated using different conditions: for example, we could change the temperature or the time taken for a particular step, the solvent, or solvent mixture, and perhaps the reagents. These experimental variables are called factors and even quite a simple chemical synthesis may involve a number of factors. What is the best way to set about altering these factors to achieve the desired goal of synthesizing the compound? We could try 'trial and error', and indeed many people do, but this is unlikely to be the most efficient way of investigating the effect of these factors, unless we are lucky. However, the most important feature of experimental design lies in the difference between 'population' values and 'sample' values. As will be described in the next chapter, any experimental result, whether a measurement or the yield from a synthesis, comes from a population of such results. When we do an experiment we wish to know about the population structure (values) using a sample to give some idea of population behaviour. In general, the larger the number of samples obtained, the better our idea of population values. The advantages of well-designed experiments are that the information can be obtained with minimum sample sizes and that the results can be interpreted to give the population information required. The next section gives some examples of strategies for experimental design. This can be of use directly in the planning of experiments but will also introduce some concepts which are of considerable importance in the analysis of property–activity data.

One may ask the question 'how is experimental design relevant to the analysis of biological data when the experimental determinations have already been made?'. One of the factors which is important in the testing of a set of compounds, and indeed intended to be the most important, is the nature of the compounds used. This set of compounds is called a training set, and selection of an appropriate training set will help to ensure that the optimum information is extracted from the experimental measurements made on the set. As will be shown in Section 2.3, the choice of training set may also determine the most appropriate physicochemical descriptors to use in the analysis of experimental data for the set. At the risk of stating the obvious, it should be pointed out that the application of any analytical method to a training set can only extract as much information as the set contains. Careful selection of the training set can help to ensure that the information it contains is maximized.

2.2 Experimental design techniques

Before discussing the techniques of experimental design it is necessary to introduce some terms which describe the important features of experiments. As mentioned in the previous section, the variables which determine the outcome of an experiment are called **factors**. Factors may be qualitative or quantitative. As an example, consider an experiment which is intended to assess how well a compound or set of compounds acts as an inhibitor of an enzyme *in vitro*. The enzyme assay will be carried out at a certain temperature and pH using a particular buffer with a given substrate and perhaps cofactor at fixed concentrations. Different buffers may be employed, as might different substrates if the enzyme catalyses a class of reaction (e.g. angiotensin converting enzyme splits off dipeptides from the C-terminal end of peptides with widely varying terminal amino acid sequences). These are qualitative factors since to change them is an 'all or nothing' change. The other factors such as temperature, pH, and the concentration of reagents are quantitative; for quantitative factors it is necessary to decide the levels which they can adopt. Most enzymes carry out their catalytic function best at a particular pH and temperature, and will cease to function at all if the conditions are changed too far from this optimum. In the case of human enzymes, for example, the optimum temperature is likely to be 37°C and the range of temperature over which they catalyse reactions may be (say) 32 to 42°C. Thus we may choose three levels for this factor: low, medium, and high, corresponding to 32, 37, and 42°C. The reason for choosing a small number of discrete levels for a continuous variable such as this is to reduce the number of possible experiments (as will be seen below). In the case of an enzyme assay, experience might lead us to expect that medium would give the highest turnover of substrate although experimental convenience might prompt the use of a different level of this factor.*

A particular set of experimental conditions is known as a **treatment** and for any experiment there are as many possible treatments as the product of the levels of each of the factors involved. Suppose that we wish to investigate the performance of an enzyme with respect to temperature, pH, and the presence or absence of a natural cofactor. The substrate concentration might be fixed at its physiological level and we might choose two levels of pH which we expect to bracket the optimum pH. Here the cofactor is a qualitative factor which can adopt one of two levels, present or absent, temperature may take three levels as before, and pH has two levels, thus there are $2 \times 3 \times 2 = 12$ possible treatments, as shown in Table 2.1. The outcome of an experiment for a given treatment is termed a **response**; in this enzyme example the response might be the rate of conversion of substrate, and in our previous example the response might be the percentage yield of compound synthesized. How can we tell the importance of the effect of a given factor on a response and how can we tell if this apparent effect is real? For example, the effect may be a population property rather than a sample property due to random variation. This can be achieved by replication—the larger the number of replicates of a given

Table 2.1 Experimental factors for an enzyme assay

Factor	Type	Levels	Treatments
Temperature (°C)	Quantitative	32, 37, 42	3
Cofactor	Qualitative	Yes/No	2
pH	Quantitative	7.0, 7.8	2
Total			12

* Optimum enzyme performance might deplete the substrate too rapidly and give an inaccurate measure of a compound as an inhibitor.

treatment then the better will be our estimate of the variation in response for that treatment. We will also have greater confidence that any one result obtained is not spurious since we can compare it with the others and thus compare variation due to the treatment to random variation. Replication, however, consumes resources such as time and material, and so an important feature of experimental design is to balance the effort between replication and change in treatment. A balanced design is one in which the treatments to be compared are replicated the same number of times, and this is desirable because it maintains orthogonality between factors (an important assumption in the analysis of variance).

The factors which have been discussed so far are susceptible to change by the experimentalist and are thus referred to as controlled factors. Other factors may also affect the experimental response and these are referred to as uncontrolled factors. How can experiments be designed to detect, and hopefully eliminate, the effects of uncontrolled factors on the response? Uncontrolled factors may very often be time-dependent. In the example of the enzyme assay, the substrate concentration may be monitored using an instrument such as a spectrophotometer. The response of the instrument may change with time and this might be confused with effects due to the different treatments unless steps are taken to avoid this. One approach might be to calibrate the instrument at regular intervals with a standard solution: calibration is, of course, a routine procedure. However, this approach might fail if the standard solution were subject to change with time, unless fresh solutions were made for each calibration. Even if the more obvious time-dependent uncontrolled factors such as instrument drift are accounted for, there may be other important factors at work.

One way to help eliminate the effect of uncontrolled factors is to randomize the order in which the different treatments are applied. The consideration that the order in which experiments are carried out is important introduces the concept of batches, known as **blocks**, of experiments. Since an individual experiment takes a certain amount of time and will require a given amount of material it may not be possible to carry out all of the required treatments on the same day or with the same batch of reagents. If the enzyme assay takes one hour to complete, it may not be possible to examine more than six treatments in a day. Taking just the factor pH and considering three levels, low (7.2), medium (7.4), and high (7.6), labelled as A, B, and C, a randomized block design with two replicates might be

A, B, B, C, A, C

Another asay might take less time and allow eight treatments to be carried out in one day. This block of experiments would enable us to examine the effect of two factors at two levels with two replicates. Taking the factors pH and cofactor, labelled as A and B for high and low levels of pH, and 1 and 0 for presence or absence of cofactor, a randomized block design with two replicates might be

A1, B0, B1, A0, A1, B0, B1, A0

This has the advantage that the presence or absence of cofactor alternates between treatments but has the disadvantage that the high pH treatments with cofactor occur at the beginning and in the middle of the block. If an instrument is switched on at the beginning of the day and then again half-way through the day, say after a lunch break, then the replicates of this particular treatment will be subject to a unique set of conditions—the one-hour warm-up period of the instrument. Similarly the low pH treatments are carried out at the same times after the instrument is switched on. This particular set of eight treatments might be better split into two blocks of four; in order to keep blocks of experiments homogeneous it pays to keep them as small as possible. Alternatively, better randomization within the block of eight treatments would help to guard against uncontrolled factors. Once again, balance is important—it may be better to examine the effect of one factor in a block of experiments using a larger number of replicates. This is the way that block designs are usually employed, examining the effect of one experimental factor while holding other factors constant. This does introduce the added complication of possible differences between the blocks. In a blocked design, the effect of a factor is of interest, not normally the effect of the blocks, so the solution is to ensure good randomization within the block and/or to repeat the block of experimental treatments.

A summary of the terms introduced so far is shown in Table 2.2.

Table 2.2 Terms used in experimental design

Term	Meaning
Response	The outcome of an experiment
Factor	A variable which affects the experimental response. These can be controlled and uncontrolled, qualitative and quantitative
Level	The values which a factor can adopt. In the case of a qualitative factor these are usually binary (e.g. present/absent)
Treatment	Conditions for a given experiment, e.g. temperature, pH, reagent concentration, solvent
Block	Set of experimental treatments carried out in a particular time-period or with a particular batch of material and thus (hopefully) under the same conditions. Generally, observations within a block can be compared with greater precision than between blocks
Randomization	Random ordering of the treatments within a block in an attempt to minimize the effect of uncontrolled factors
Replication	Repeat experimental treatments to estimate the significance of the effect of individual factors on the response (and to identify 'unusual' effects)
Balance	The relationship between the number of treatments to be compared and the number of replicates of each treatment examined

2.2.1 Single-factor design methods

The block design shown previously is referred to as a 'balanced, complete block design' since all of the treatments were examined within the block (hence 'complete'), and the number of replicates was the same for all treatments (hence 'balanced'). If the number of treatments and their replicates is larger than the number of experimental 'slots' in a block then it will be necessary to carry out two or more blocks of experiments to examine the effect of the factor. This requires that the blocks of experiments are chosen in such a way that comparisons between treatments will not be affected. When all of the comparisons are of equal importance (for example, low vs. high temperature, low vs. medium, and high vs. medium) the treatments should be selected in a balanced way so that any two occur together the same number of times as any other two. This type of experimental design is known as 'balanced, incomplete block design'. The results of this type of design are more difficult to analyse than the results of a complete design, but easier than if the treatments were chosen at random which would be an 'unbalanced, incomplete block design'.

The time taken for an individual experiment may determine how many experiments can be carried out in a block, as may the amount of material required for each treatment. If both of these factors, or any other two 'blocking variables', are important then it is necessary to organize the treatments to take account of two (potential) uncontrolled factors. Suppose that: there are three possible treatments, A, B, and C; it is only possible to examine three treatments in a day; a given batch of material is sufficient for three treatments; time of day is considered to be an important factor. A randomized design for this is shown below.

Batch	Time of day		
1	A	B	C
2	B	C	A
3	C	A	B

This is known as a **Latin square**, perhaps the best-known term in experimental design, and is used to ensure that the treatments are randomized to avoid trends within the design. Thus, the Latin square design is used when considering the effect of one factor and two blocking variables. In this case the factor was divided into three levels giving rise to three treatments: this requires a 3×3 matrix. If the factor has more levels, then the design will simply be a larger symmetrical matrix, i.e. 4×4, 5×5, and so on. What about the situation where there are three blocking variables? In the enzyme assay example, time of day may be important and there may only be sufficient cofactor in one batch for three assays and similarly only sufficient enzyme in one batch for three assays. This calls for a design known as **Graeco-Latin square** which is made by superimposing two different Latin squares. There are two possible 3 × 3 Latin squares:

A	B	C
B	C	A
C	A	B

A	B	C
C	A	B
B	C	A

The 3 × 3 Graeco-Latin square is made by the superimposition of these two Latin squares with the third blocking variable denoted by Greek letters thus:

Aα	Bβ	Cγ
Bγ	Cα	Aβ
Cβ	Aγ	Bα

It can be seen in this design that each treatment occurs only once in each row and column (two of the blocking variables, say time of day and cofactor batch) and only once with each level (α, β, and γ) of the third blocking variable, the enzyme batch. Both Latin squares and Graeco-Latin squares (and **Hyper-Graeco-Latin squares** for more blocking variables) are most effective if they are replicated and are also subject to the rules of randomization which apply to simple block designs. While these designs are useful in situations where only one experimental factor is varied, it is clear that if several factors are important (a more usual situation), this approach will require a large number of experiments to examine their effects. Another disadvantage of designing experiments to investigate a single factor at a time is that the interactions between factors are not examined since in this approach all other factors are kept constant.

2.2.2 Factorial design (multiple-factor design)

The simplest example of the consideration of multiple experimental factors would involve two factors. Taking the earlier example of a chemical synthesis, suppose that we were interested in the effect of two different reaction temperatures, T_1 and T_2, and two different solvents, S_1 and S_2, on the yield of the reaction. The minimum number of experiments required to give us information on both factors is three, one at T_1S_1 (y_1), a second at T_1S_2 (y_2) involving change in solvent, and a third at T_2S_1 (y_3) involving a change in temperature (see Table 2.3). The effect of changing temperature

Table 2.3 Options in a multiple-factor design

Solvent	Temperature	
	T_1	T_2
S_1	y_1	y_3
S_2	y_2	y_4

is given by the difference in yields $y_3 - y_1$ and the effect of changing solvent is given by $y_2 - y_1$. Confirmation of these results could be obtained by duplication of the above requiring a total of six experiments. This is a 'one variable at a time' approach since each factor is examined separately. However, if a fourth experiment, T_2S_2 (y_4), is added to Table 2.3 we now have two measures of the effect of changing each factor but only require four experiments. In addition to saving two experimental determinations, this approach allows the detection of interaction effects between the factors, such as the effect of changing temperature in solvent 2 ($y_4 - y_2$) compared with solvent 1 ($y_3 - y_1$). The factorial approach is not only more efficient in terms of the number of experiments required and the identification of interaction effects, it can also be useful in optimization. For example, having estimated the main effects and interaction terms of some experimental factors it may be possible to predict the likely combinations of these factors which will give an optimum response. One drawback to this procedure is that it may not always be possible to establish all possible combinations of treatments, resulting in an unbalanced design. Factorial designs also tend to involve a large number of experiments, the investigation of three factors at three levels, for example, requires 27 runs (3^f where f is the number of factors) without replication of any of the combinations. However, it is possible to reduce the number of experiments required as will be shown later.

A recently published example (Coleman *et al.* 1993) nicely illustrates the use of factorial design in chemical synthesis. The reaction of 1,1,1-trichloro-3-methyl-3-phospholene (**1**) with methanol produces 1-methoxy-3-methyl-2-phospholene oxide (**2**) as shown in the reaction scheme. The experimental procedure involved the slow addition of a known quantity of

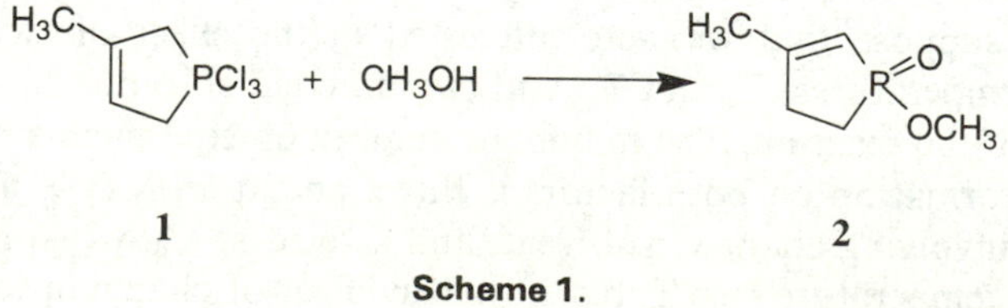

H_3C — PCl_3 (**1**) + CH_3OH → H_3C — $P(=O)OCH_3$ (**2**)

Scheme 1.

methanol to a known quantity of **1** in dichloromethane held at subambient temperature. The mixture was then stirred until it reached ambient temperature and neutralized with aqueous sodium carbonate solution; the product was extracted with dichloromethane. The yield from this reaction was 25 per cent and could not significantly be improved by changing one variable (concentration, temperature, addition time, etc.) at a time. Three variables were chosen for investigation by factorial design using two levels of each.

A: Addition temperature (−15 or 0°C)
B: Concentration of **1** (50 or 100g in 400cm^3 dichloromethane)
C: Addition time of methanol (one or four hours)

This led to eight different treatments (2^3), which resulted in several yields above 25 per cent (as shown in Table 2.4), the largest being 42.5 per cent.

The effect on an experimental response due to a factor is called a **main** effect whereas the effect caused by one factor at each level of the other factor is called an **interaction** effect (two way). The larger the number of levels of thc factors studied in a factorial design, the higher the order of the interaction effects that can be identified. In a three-level factorial design it is possible to detect quadratic effects although it is often difficult to interpret the information. Three-level factorial designs also require a considerable number of experiments (3^f) as shown above. For this reason it is often found convenient to consider factors at just two levels, high/low or yes/no, to give 2^f factorial designs.

Another feature of these full factorial designs, full in the sense that all combinations of all levels of each factor are considered, is that interactions between multiple factors may be identified. In a factorial design with six

Table 2.4 Responses from full factorial design (from Coleman *et al.* 1993, with permission of the Royal Society of Chemistry)

Order of treatment	Treatment combination[a]	Yield (%)
3	–	24.8
6	a	42.5
1	b	39.0
7	ab	18.2
2	c	32.8
4	ac	33.0
8	bc	13.2
5	abc	24.3

[a] Where a lower-case letter is shown, this indicates that a particular factor was used at its high level in that treatment, e.g. 'a' means an addition temperature of 0°C. When a letter is missing the factor was at its low level.

factors at two levels ($2^6 = 64$ experiments) there are six main effects (for the six factors), 15 two-factor interactions (two-way effects), 20 three-factor, 15 four-factor, 6 five-factor, and 1 six-factor interactions. Are these interactions all likely to be important? The answer, fortunately, is no. In general, main effects tend to be larger than two-factor interactions which in turn tend to be larger than three-factor interactions and so on. Because these higher order interaction terms tend not to be significant it is possible to devise smaller factorial designs which will still investigate the experimental factor space efficiently but which will require far fewer experiments. It is also often found that in factorial designs with many experimental factors, only a few factors are important. These smaller factorial designs are referred to as **fractional factorial designs**, where the fraction is defined as the ratio of the number of experimental runs needed in a full design. For example, the full factorial design for five factors at two levels requires 32 (2^5) runs: if this is investigated in 16 experiments it is a half-fraction factorial design. Fractional designs may also be designated as 2^{f-n} where f is the number of factors as before and n is the number of half-fractions, 2^{5-1} is a half-fraction factorial design in five factors, 2^{6-2} is a quarter-fraction design in six factors.

Of course, it is rare in life to get something for nothing and that principle applies to fractional factorial designs. Although a fractional design allows one to investigate an experimental system with the expenditure of less effort, it is achieved at the expense of clarity in our ability to separate main effects from interactions. The response obtained from certain treatments could be caused by the main effect of one factor or a two- (three-, four-, five-, etc.) factor interaction. These effects are said to be confounded; because they are indistinguishable from one another, they are also said to be aliases of one another. It is the choice of aliases which lies at the heart of successful fractional factorial design. As mentioned before, we might expect that main effects would be more significant than two-factor effects which will be more important than three-factor effects. The aim of fractional design is thus to alias main effects and two-factor effects with as high-order interaction terms as possible.

The phospholene oxide synthesis mentioned earlier provides a good example of the use of fractional factorial design. Having carried out the full factorial design in three factors (addition temperature, concentration of phospholene, and addition time) further experiments were made to 'fine-tune' the response. These probing experiments involved small changes to one factor while the others were held constant in order to determine whether an optimum had been reached in the synthetic conditions. Figure 2.1 shows a response surface for the high addition time in which percentage yield is plotted against phospholene concentration and addition temperature. The response surface is quite complex and demonstrates that a maximum yield had not been achieved for the factors examined in the

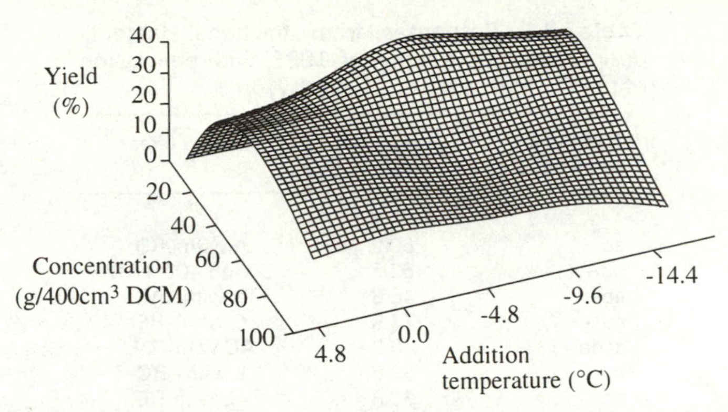

Fig. 2.1. Response surface for phospholene oxide synthesis (from Coleman *et al.* 1993, with permission of the Royal Society of Chemistry).

first full factorial design. In fact the largest yield found in these probing experiments was 57 per cent, a reasonable increase over the highest yield of 42.5 per cent shown in Table 2.4. The shape of the response surface suggests the involvement of other factors in the yield of this reaction and three more experimental variables were identified: concentration of methanol, stirring time, and temperature. Fixing the concentration of phospholene at 25 g in 400 cm^3 of dichloromethane (a broad peak on the response surface) leaves five experimental factors to consider, requiring a total of 32 (2^5) experiments to investigate them. These experiments were split into four blocks of eight and hence each block is a quarter-fraction of 32 experiments. The results for the first block are shown in Table 2.5, the experimental factors being

A: Addition temperature (−10 or 0°C)
B: Addition time of methanol (15 or 30 minutes)
C: Concentration of methanol (136 or 272 cm^3)
D: Stirring time (0.5 or 2 hours)
E: Stirring temperature (addition temperature or ambient)

This particular block of eight runs was generated by aliasing D with AB and also E with BC, after carrying out full 2^3 experiments of A, B, and C. As can be seen from Table 2.5, the best yield from this principal block of experiments, which contains variables and variable interactions expected to be important, was 78 per cent, a considerable improvement over the previously found best yield of 57 per cent. Having identified important factors, or combinations of factors with which they are aliased, it is possible to choose other treatment combinations which will clarify the situation. The best yield obtained for this synthesis was 90 per cent using treatment combination e.

Table 2.5 Responses from fractional factorial design (from Coleman *et al.* 1993, with permission of the Royal Society of Chemistry)

Treatment combination[a]	Yield (%)	Aliasing effect
–	45.1	
ad	60.2	A with –BD
bde	62.5	B with –CE +AD
abe	46.8	D with –AB
ce	77.8	C with –BE
acde	49.8	AC with –DE
bcd	53.6	E with –BC
abc	70.8	AE with CD

[a] As explained in Table 2.4.

2.2.3 D-optimal design

Factorial design methods offer the advantage of a systematic exploration of the factors that are likely to affect the outcome of an experiment; they also allow the identification of interactions between these factors. They suffer from the disadvantage that they may require a large number of experiments, particularly if several levels of each factor are to be examined. This can be overcome to some extent by the use of fractional factorial designs although the aliasing of multi-factor interactions with main effects can be a disadvantage. Perhaps one of the biggest disadvantages of factorial and fractional factorial designs in chemistry is the need to specify different levels for the factors. If a factorial design approach is to be successful, it is necessary to construct treatment combinations which explore the factor space. When the experimental factors are variables such as time, temperature, and concentration, this usually presents few problems. However, if the factors are related to chemical structure, such as the choice of compounds for testing, the situation may be quite different. A factorial design may require a particular compound which is very difficult to synthesize. Alternatively, a design may call for a particular set of physicochemical properties which cannot be achieved, such as a very small, very hydrophobic substituent.

The philosophy of the factorial approach is attractive, so are there related techniques which are more appropriate to the special requirements of chemistry? There is a number of other methods for experimental design but one that is becoming applied in several chemical applications is known as 'D-optimal design'. The origin of the expression 'D-optimal' is a bit of statistical jargon based on the determinant of the variance–covariance matrix. As will be seen in the next section on compound selection, a well-chosen set of experiments (or compounds) will have a wide spread in the experimental factor space (variance). A well-chosen set will also be such that the correlation (see Box 2.1) between experimental factors is at a minimum (covariance). The determinant of a matrix is a single number and in the case of a variance–covariance matrix for a data set this number describes the 'balance' between variance and covariance. This determinant will be a maximum for experiments or compounds which have maximum variance and minimum covariance, and thus the optimization of the determinant (D-optimal) is the basis of the design. Examples of the use of D-optimal design are given in the next section.

Further discussion of other experimental design techniques, with the exception of **simplex** optimization (see next section), is outside the scope of this book. Hopefully this section will have introduced the principles of experimental design; the reader interested in further details should consult one of the excellent texts available which deal with this subject in detail (see Box *et al.* 1978; Morgan 1991). A recent review discusses the application of experimental design techniques to chemical synthesis (Carlson and Nordahl 1993).

2.3 Strategies for compound selection

This section could also have been entitled strategies for 'training set selection' where compounds are the members of a training set. Training sets are required whenever a new biological test is established, when compounds are selected from an archive for screening in an existing test, or when a set of biological (or other) data is to be analysed. It cannot be stressed sufficiently that selection of appropriate training sets is crucial to the success of new synthetic programmes, screening, and analysis. The following examples illustrate various aspects of compound selection.

Some of the earliest techniques for compound selection were essentially visual and as such have considerable appeal compared with the (apparently) more complex statistical and mathematical methods. The first method to be reported came from a study of the relationships between a set of commonly used substituent constants (Craig 1971). The stated purpose of this work was to examine the interdependence of these parameters and, as expected, correlations (see Box 2.1) were found between the hydrophobicity descriptor, π, and a number of 'bulk' parameters such as molecular volume and parachor. Why should interdependence between substituent constants be important? There are a number of answers to this question, as discussed further in this book, but for the present it is sufficient to say that interdependence between parameters is required so that clearer, perhaps mechanistic, conclusions might be drawn from correlations. As part of the investigation Craig plotted various parameters together, for example the plot of σ vs. π shown in Fig. 2.2; such plots have

Box 2.1

The correlation coefficient, *r*

An important property of any variable, which is used in many statistical operations is a quantity called the variance, V. The variance is a measure of how the values of a variable are distributed about the mean and is defined by

$$V = \sum_{i=1}^{n} (x_i - \bar{x})^2 / n$$

where $\bar{x}$ is the mean of the set of x values and the summation is carried out over all n members of the set. When values are available for two or more variables describing a set of objects (compounds, samples, etc.) a related quantity may be calculated called the covariance, $C_{(x,y)}$. The covariance is a measure of how the values of one variable (x) are distributed about their mean compared with how the corresponding values of another variable (y) are distributed about their mean. Covariance is defined as

$$C_{(x,y)} = \sum_{i=1}^{n} (x_i - \bar{x})(y_i - \bar{y}) / n$$

The covariance has some useful properties. If the values of variable x change in the same way as the values of variable y, the covariance will be positive. For small values of x, $x - \bar{x}$ will be negative and $y - \bar{y}$ will be negative yielding a positive product. For large values of x, $x - \bar{x}$ will be positive as will $y - \bar{y}$ and thus the summation yields a positive number. If, on the other hand, y decreases as x increases the covariance will be negative. The sign of the covariance of the two variables indicates how they change with respect to one another: positive if they go up and down together, negative if one increases as the other decreases. But is it possible to say how clearly one variable mirrors the change in another? The answer is yes, by the calculation of a quantity known as the correlation coefficient

$$r = C_{(x,y)} / [V_{(x)} \times V_{(y)}]^{\frac{1}{2}}$$

Division of the covariance by the square root of the product of the individual variances allows us to put a scale on the degree to which two variables are related. If y changes by exactly the same amount as x changes, and in the same direction, the correlation coefficient is $+1$. If y decreases by exactly the same amount as x increases the correlation coefficient is -1. If the changes in y are completely unrelated to the changes in x, the correlation coefficient will be 0.

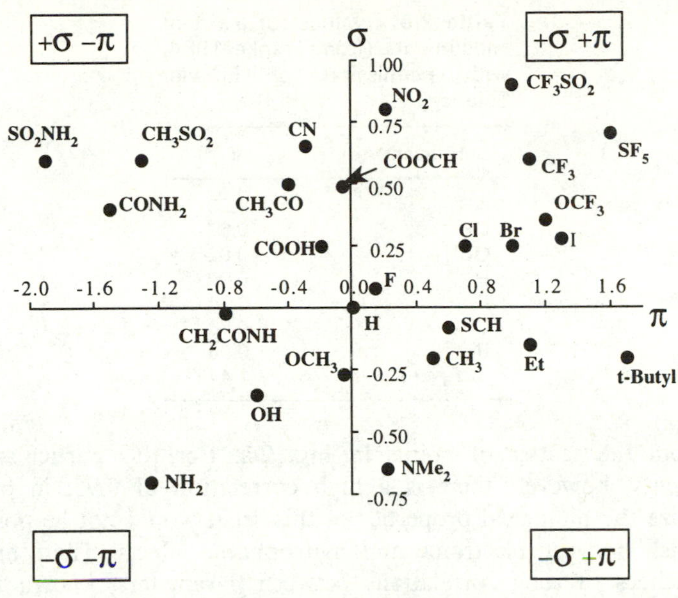

Fig. 2.2. Plot of σ vs π for a set of common substituents (from Craig 1971, copyright (1971) American Chemical Society).

since become known as Craig plots. This diagram nicely illustrates the concept of a physicochemical parameter space. If we regard these two properties as potentially important experimental factors, in the sense that they are likely to control or at least influence experiments carried out using the compounds, then we should seek to choose substituents that span the parameter space. This is equivalent to the choice of experimental treatments which are intended to span the space of the experimental factors.

It is easy to see how substituents may be selected from a plot such as that shown in Fig. 2.2, but will this ensure that the series is well chosen? The answer is no for two reasons. First, the choice of compounds based on the parameter space defined by just two substituent constants ignores the potential importance of any other factors. What is required is the selection of points in a multidimensional space, where each dimension corresponds to a physicochemical parameter, so that the space is sampled evenly. This is described later in this section. The second problem with compound choice based on sampling a two-parameter space concerns the correlation between the parameters. Table 2.6 lists a set of substituents with their corresponding π values. At first sight this might appear to be a well-chosen set since the substituents cover a range of -1.2 to $+1.4$ log units in π and are represented at fairly even steps over the range. If, however, we now list the σ values for these substituents, as shown in Table 2.7, we see that they also span a good range of σ but that the two sets of values correspond to one another. In general, there is no correlation between π and σ as can be

Table 2.6 π values for a set of substituents (from Franke 1984, with permission of Elsevier Science)

Substituent	π
NH_2	−1.23
OH	−0.67
OCH_3	−0.02
H	0.00
F	0.14
Cl	0.70
Br	0.86
SCF_3	1.44

seen from the scatter of points in Fig. 2.2. For this particular set of substituents, however, there is a high correlation of 0.95; in trying to rationalize the biological properties of this set it would not be possible to distinguish between electronic and hydrophobic effects. There are other consequences of such correlations between parameters, known as collinearity, which involve multiple regression (Chapter 6), data display (Chapter 4), and other multivariate methods (Chapters 7 and 8). This is discussed in the next chapter and in the chapters which detail the techniques.

Table 2.7 π and σ values for the substituents in Table 2.6 (from Franke 1984, with permission of Elsevier Science)

Substituent	π	σ
NH_2	−1.23	−0.66
OH	−0.67	−0.37
OCH_3	−0.02	−0.27
H	0.00	0.00
F	0.14	0.06
Cl	0.70	0.23
Br	0.86	0.23
SCF_3	1.44	0.50

So, the two main problems in compound selection are the choice of analogues to sample effectively a multi-parameter space and the avoidance of collinearity between physicochemical descriptors. A number of methods have been proposed to deal with these two problems. An attractive approach was published by Hansch and co-workers (Hansch *et al.* 1973) which made use of cluster analysis (Chapter 5) to group 90 substituents described by five physicochemical parameters. Briefly, cluster analysis operates by the use of measurements of the distances between pairs of objects in multidimensional space using a distance such as the familiar Euclidean distance. Objects (compounds) which are close together in space become members of a single cluster. For a given level of similarity (i.e. value of the distance measure) a given number of clusters will be formed for a particular data set. At decreasing levels of similarity (greater values of the distance measure) further objects or clusters will be joined to the original clusters until eventually all objects in the set belong to a single cluster. The results of cluster analysis are most often reported in the form of a diagram known as a dendrogram (Fig. 2.3). A given level of similarity on the dendrogram gives rise to a particular number of clusters and thus it was possible for Hansch and his co-workers to produce lists of substituents belonging to 5, 10, 20, and 60 cluster sets. This allows a medicinal chemist to choose a substituent from each cluster when making a particular number of training set compounds (5, 10, 20, or 60) to help ensure that parameter space is well spanned. This work was subsequently updated to cover 166 substituents described by six parameters; lists of the cluster members were reported in the substituent constant book (Hansch and Leo 1978), sadly no longer in print. Table 2.8 lists some of the substituent members of the ten cluster set.

Another approach which makes use of the distances between points in a multidimensional space was published by Wootton and co-workers (Wootton *et al.* 1975). In this method the distances between each pair of substituents is calculated, as described for cluster analysis, and substituents are chosen in a stepwise fashion such that they exceed a certain preset minimum distance. The procedure requires the choice of a particular starting compound, probably but not necessarily the unsubstituted parent,

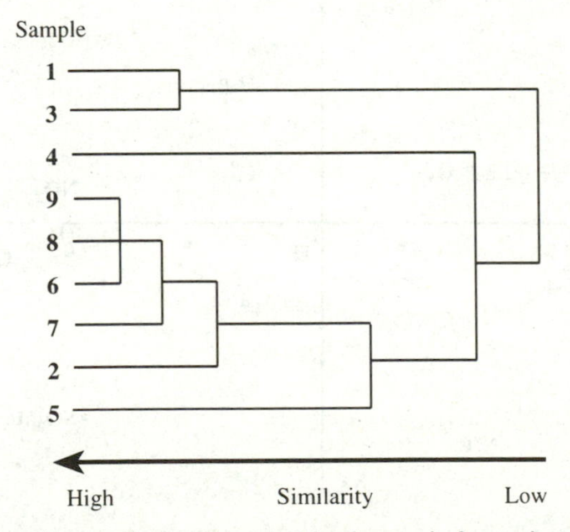

Fig. 2.3. Example of a similarity diagram (dendrogram) from cluster analysis (after Reibnegger *et al.* 1993, with permission of Butterworth–Heinemann).

Table 2.8 Examples of substituents belonging to clusters in the ten cluster set (from Hansch and Leo 1979, with their kind permission)

Cluster set number	Number of members	Examples of substituents
1	26	–Br, –Cl, –NNN, $–CH_3$, $–CH_2Br$
2	17	$–SO_2F$, $–NO_2$, –CN, –1–Tetrazolyl, $–SOCH_3$
3	2	$–IO_2$, $–N(CH_3)_3$
4	8	–OH, $–NH_2$, $–NHCH_3$, $–NHC_4H_9$, $–NHC_6H_5$
5	18	$–CH_2OH$, –NHCN, $–NHCOCH_3$, $–CO_2H$, $–CONH_2$
6	21	$–OCF_3$, $–CH_2CN$, –SCN, $–CO_2CH_3$, –CHO
7	25	–NCS, –Pyrryl, $–OCOC_3H_7$, $–COC_6H_5$, $–OC_6H_5$
8	20	$–CH_2I$, $–C_6H_5$, $–C_5H_{11}$, –Cyclohexyl, $–C_4H_9$
9	21	$–NHC{=}S(NH_2)$, $–CONHC_3H_7$, $–NHCOC_2H_5$, $–C(OH)(CF_3)_2$, $–NHSO_2C_6H_5$
10	8	$–OC_4H_9$, $–N(CH_3)_2$, $–N(C_2H_5)_2$

and choice of the minimum distance. Figure 2.4 gives an illustration of this process to the choice of eight substituents from a set of 35. The resulting correlation between the two parameters for this set was low (−0.05). A related technique has been described by Franke (Streich *et al.* 1980) in which principal components are calculated from the physicochemical descriptor data (see Chapter 4) and interpoint distances are calculated based on the principal components. Several techniques are compared in the reference cited.

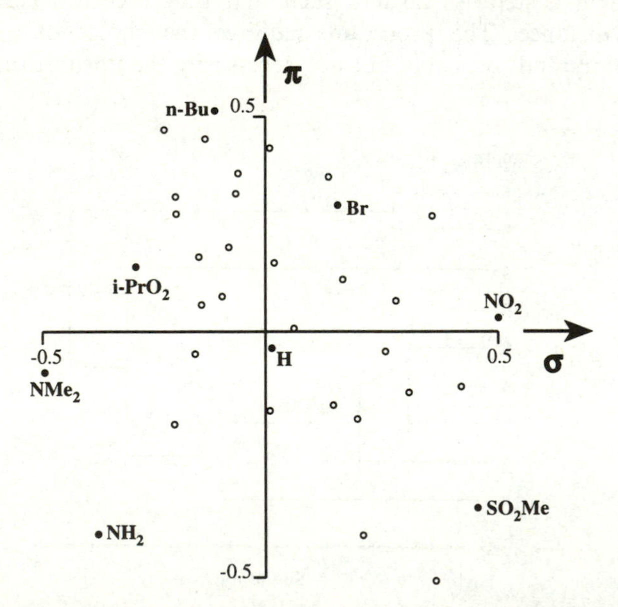

Fig. 2.4. Example of the choice of substituents by multidimensional mapping (from Wootton *et al.* 1975, copyright (1975) American Chemical Society).

These techniques for compound selection have relied on the choice of substituents such that the physicochemical parameter space is well covered; the resulting sets of compounds tend to be well spread and interparameter correlations low. These were the two criteria set out earlier for successful compound choice, although other criteria, such as synthetic feasibility, may be considered important (Schaper 1983). An alternative way to deal with the problem of compound selection is to treat the physicochemical properties as experimental factors and apply the techniques of factorial design. As described in Section 2.2, it is necessary to decide how many levels need to be considered for each individual factor in order to determine how many experimental treatments are required. Since the number of experiments (and hence compounds) increases as the product of the factor levels, it is usual to consider just two levels, say high and low, for each factor. This also allows qualitative factors such as the presence/absence of some functional group or structural feature to be included in the design. Of course, if some particular property is known or suspected to be of importance, then this may be considered at more than two levels. A major advantage of factorial design is that many factors may be considered at once and that interactions between factors may be identified, unlike the two parameter treatment of Craig plots. A disadvantage of factorial design is the large number of experiments that may need to be considered, but this may be reduced by the use of fractional factorials as described in Section 2.2. Austel (1982) was the first to describe factorial designs for compound selection and he demonstrated the utility of this approach by application to literature examples. The relationship of a full to a half-fractional design is nicely illustrated in Fig. 2.5. The cube represents the space defined by three physicochemical properties A, B, and C and the points at the vertices represent the compounds chosen to examine various combinations of these parameters as shown in Table 2.9. An extra point can usefully be considered in designs such as this corresponding to

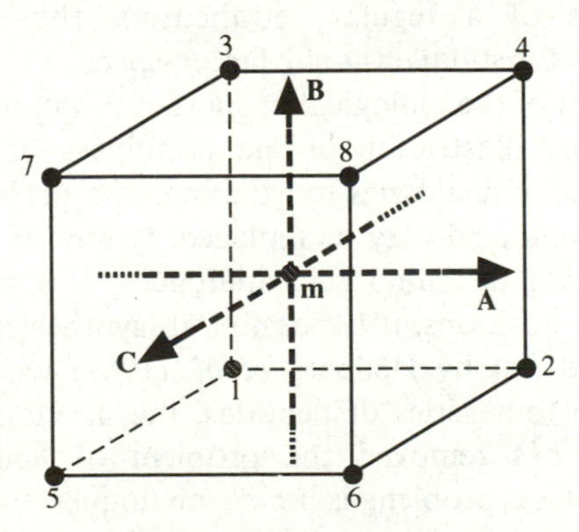

Fig. 2.5. Representation of a factorial design in three factors (A, B, and C) (from Austel 1982, with permission of the *European Journal of Medicinal Chemistry*).

Table 2.9 Factorial design for three parameters (two levels) (from Austel 1982, with permission of the *European Journal of Medicinal Chemistry*)

Compound	A	B	C
1	−	−	−
2	+	−	−
3	−	+	−
4	+	+	−
5	−	−	+
6	+	−	+
7	−	+	+
8	+	+	+

Table 2.10 Fractional factorial design for three parameters (two levels) (from Austel 1982, with permission of the *European Journal of Medicinal Chemistry*)

Compound	A	B	C
1(5)[a]	−	−	+
2(2)	+	−	−
3(3)	−	+	−
4(8)	+	+	+

[a] ( ) corresponding compound in Table 2.9.

the midpoint of the factor space. If A, B, and C are substituent constants such as π, σ, and MR which are scaled to H = 0, this midpoint is the unsubstituted parent. A fractional factorial design in these three parameters is shown in Table 2.10. This fractional design investigates the main effects of parameters A and B, factor C is confounded (aliased, see Section 2.2.2) with interaction of A and B. The four compounds in this table correspond to compounds 2, 3, 5, and 8 from Table 2.9 and in Fig. 2.5 form the vertices of a regular tetrahedron, thus providing a good exploration of the three-dimensional factor space.

The investigation of the biological properties of peptide analogues gives a particularly striking illustration of the usefulness of fractional factorial design in the choice of analogues to examine. The problem with peptides is that any single amino acid may be replaced by any of the 20 coded amino acids, to say nothing of amino acid analogues. If a peptide of interest is varied in just four positions, it is possible to synthesize 160,000 (20^4) analogues. As pointed out by Hellberg *et al.* (1987) who applied fractional factorial design to four series of peptides, the development of automated peptide synthesis has removed the problem of *how* to make peptide analogues. The major problem is *which* analogues to make. In order to apply the principles of experimental design to this problem it is necessary to define experimental factors (physicochemical properties) to be explored.

Table 2.11 Descriptor scales for the 20 'natural' amino acids (from Hellberg *et al.* 1987, copyright (1987) American Chemical Society)

Acid	Z_1	Z_2	Z_3
Ala	0.07	−1.73	0.09
Val	−2.69	−2.53	−1.29
Leu	−4.19	−1.03	−0.98
Ile	−4.44	−1.68	−1.03
Pro	−1.22	0.88	2.23
Phe	−4.92	1.30	0.45
Trp	−4.75	3.65	0.85
Met	−2.49	−0.27	−0.41
Lys	2.84	1.41	−3.14
Arg	2.88	2.52	−3.44
His	2.41	1.74	1.11
Gly	2.23	−5.36	0.30
Ser	1.96	−1.63	0.57
Thr	0.92	−2.09	−1.40
Cys	0.71	−0.97	4.13
Tyr	−1.39	2.32	0.01
Asn	3.22	1.45	0.84
Gln	2.18	0.53	−1.14
Asp	3.64	1.13	2.36
Glu	3.08	0.39	−0.07

These workers used 'principal properties' which were derived from the application of principal component analysis (see Chapter 4) to a data matrix of 29 physicochemical variables which describe the amino acids. The principal component analysis gave three new variables, labelled Z_1, Z_2, and Z_3, which were interpreted as being related to hydrophobicity (partition), bulk, and electronic properties respectively. Table 2.11 lists the values of these descriptor scales. This is very similar to an earlier treatment of physicochemical properties by Cramer (1980*a,b*), the so-called *BC(DEF)* scales. The *Z* descriptor scales thus represent a three-dimensional property space for the amino acids. If only two levels are considered for each descriptor, high (+) and low (−), a full factorial design for substitution at one amino acid position in a peptide would require eight analogues. While this is a saving compared with the 20 possible analogues that could be made, a full factorial design is impractical when multiple substitution positions are considered. A full design for four amino acid positions requires 4096 (8^4) analogues, for example. Hellberg suggested $\frac{1}{2}$ (for one position), $\frac{1}{8}$ th and smaller fractional designs as shown in Table 2.12.

One of the problems with the use of factorial methods for compound selection is that it may be difficult or impossible to obtain a compound required for a particular treatment combination, either because the synthesis is difficult or because that particular set of factors does not exist. One way to overcome this problem, as discussed in Section 2.2.3, is D-

Table 2.12 Number of peptide analogues required for fractional factorial design based on three *Z* scales (from Hellberg *et al.* 1987, copyright (1987) American Chemical Society)

Number of varied positions	Minimum number of analogues
1	4
2	8
3–5	16
6–10	32
11–21	64

optimal design. Unger (1980) has reported the application of a D-optimal design procedure to the selection of substituents from a set of 171, described by seven parameters.* The determinant of the variance–covariance matrix for the selected set of 20 substituents was 3.35×10^{11} which was 100 times better than the largest value (2.73×10^{9}) obtained in 100 simulations in which 20 substituents were randomly chosen. Herrmann (1983) has compared the use of D-optimal design, two variance maximization methods, and an information-content maximization technique for compound selection. The results of the application of these strategies to the selection of ten substituents from a set of 35 are shown in Table 2.13. Both the D-optimal and the information-content methods produced better sets of substituents, as measured by variance (V) or determinant (D) values, than the variance maximization techniques.

The final compound selection procedures which will be mentioned here are the sequential simplex and the 'Topliss tree'. The sequential simplex, first reported by Darvas (1974), is an application of a well-known optimization method which can be carried out graphically. Figure 2.6 shows three compounds, A, B, and C, plotted in a two-dimensional property space, say π and σ but any two properties may be used. Biological results are obtained for the three compounds and they are ranked in order of activity. These compounds form a triangle in the two-dimensional property space, a new compound is chosen by the construction of a new triangle. The two most active compounds, say B and C for this example, form two of the vertices of the new triangle and the third vertex is found by taking a point opposite to the least active (A) to give the new triangle BCD. The new compound is tested and the activities of the three compounds compared—if B is now the least active then a new triangle CDE is constructed as shown in the figure. The procedure can be repeated until no further improvement in activity is obtained, or until all of the attainable physicochemical property space has

* The reference includes listings of programs (written in APL) for compound selection by D-optimal design.

Table 2.13 Subsets of ten substituents from 35 chosen by four different methods (from Herrmann 1983, with permission)

Method[a]			
h	D	V(1)	V(2)
H	H	H	Me
n-butyl	*n*-butyl	phenyl	*t*-butyl
phenyl	*t*-butyl	OH	OEt
CF_3	O–phenyl	O–phenyl	O–*n*-amyl
O–phenyl	NH_2	NMe_2	NH_2
NH_2	NMe_2	NO_2	NMe_2
NMe_2	NO_2	COOEt	NO_2
NO_2	SO_2Me	$CONH_2$	$COO(Me)_2$
SO_2NH_2	SO_2NH_2	SO_2Me	SO_2NH_2
F	F	Br	F
V = 1.698	1.750	1.437	1.356
D = 1.038	1.041	0.487	0.449
h = 1.614	1.589	1.502	1.420

[a] The methods are: h, maximization of information content; D, D-optimal design; V(1), maximal variance (Streich *et al.* 1980); and V(2), maximal variance (Wootton *et al.* 1975).

been explored. This method is attractive in its simplicity and the fact that it requires no more complicated equipment than a piece of graph paper. The procedure is designed to ensure that an optimum is found in the particular parameters chosen so its success as a compound selection method is dependent on the correct choice of physicochemical properties. One of the problems with this method is that a compound may not exist that corresponds to a required simplex point. The simplex procedure is intended to operate with continuous experimental variables such as temperature, pressure, concentration, etc. There are other problems with the simplex

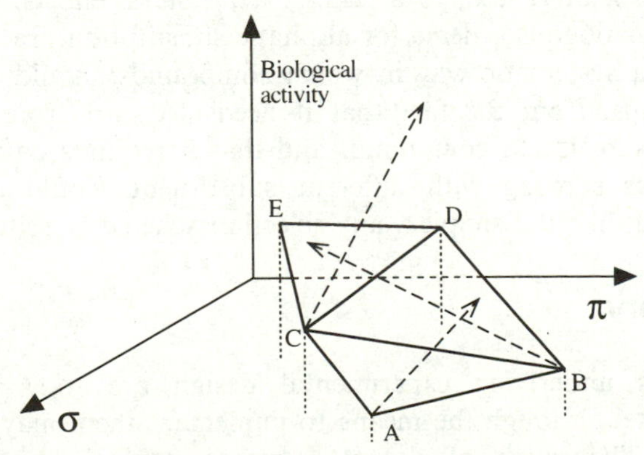

Fig. 2.6. Illustration of the sequential simplex process of compound selection (after Darvas 1974, copyright (1974) American Chemical Society).

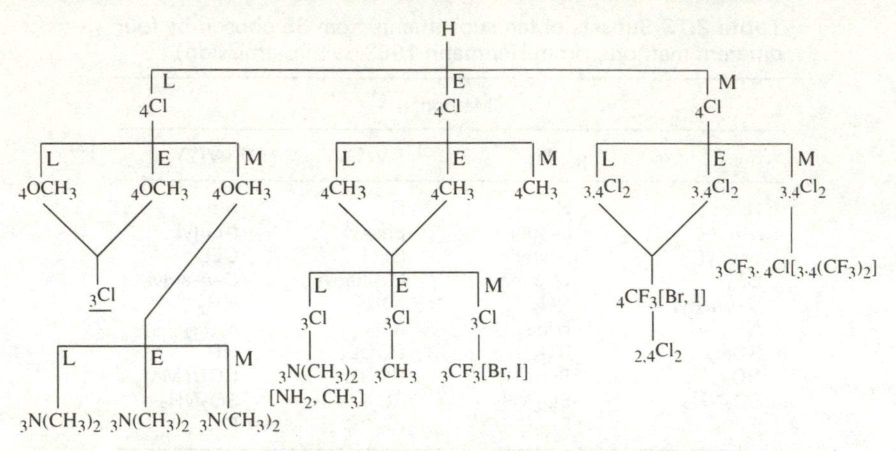

Fig. 2.7. Illustration of the 'Topliss tree' process for compound selection; L, E, and M represent less, equal, and more active respectively (after Topliss and Martin 1975, with permission of Academic Press).

procedure, for example, it requires biological activity data, but it has a number of advantages, not least of which being that *any* selection procedure is better than none.

The 'Topliss tree' is an operational scheme that is designed to explore a given physicochemical property space in an efficient manner and is thus related to the sequential simplex approach (Topliss and Martin 1975). In the case of aromatic substitution, for example, this approach assumes that the unsubstituted compound and the *para*-chloro derivative have both been made and tested. The activity of these two compounds are compared and the next substituent is chosen according to whether the chloro substituent displays higher, lower, or equal activity. This is shown schematically in Fig. 2.7. The rationale for the choice of $-OCH_3$ or Cl as the next substituent is based on the supposition that the given effects are dependent on changes in π or σ and, to a lesser extent, steric effects. This decision tree and the analogous scheme for aliphatic substitution are useful in that they suggest a systematic way in which compounds should be chosen. It suffers, perhaps, from the fact that it needs to start from a particular point, the unsubstituted compound, and that it requires data to guide it. Other schemes starting with different substituents could of course be drawn up and, like the simplex, any selection scheme is better than none.

2.4 Summary

The concepts underlying experimental design are to a great extent 'common sense' although the means to implement them may not be quite so obvious. The value of design, whether applied to an individual experiment or to the construction of a training set, should be clear from the examples shown in this chapter. Failure to apply some sort of design strategy may lead to a set of results which contain suboptimal information, at best, or which contain no useful information, at worst. Various design procedures may be applied to individual experiments, as indicated in the previous sections, and there are specialist reports which deal with topics such as synthesis (Carlson and Nordahl 1993). A detailed review of design strategies which may be applied to the selection of compounds has been reported by Pleiss and Unger (1990).

References

Austel, V. (1982). *European Journal of Medicinal Chemistry,* **17**, 9–16.

Box, G. E. P., Hunter, W. J., and Hunter, J. S. (1978). *Statistics for experimentalists.* Wiley, New York.

Carlson, R. and Nordahl, A. (1993). *Topics in Current Chemistry*, **166**, 1–64.

Coleman, G. V., Price, D., Horrocks, A. R., and Stephenson, J. E. (1993). *Journal of the Chemical Society-Perkin Transactions II*, 629–32.

Craig, P. N. (1971). *Journal of Medicinal Chemistry*, **14**, 680–4.

Cramer, R. D. (1980a). *Journal of the American Chemical Society*, **102**, 1837–49.

Cramer, R. D. (1980b). *Journal of the American Chemical Society*, **102**, 1849–59.

Darvas, F. (1974). *Journal of Medicinal Chemistry*, **17**, 799–804.

Franke, R. (1984). *Theoretical drug design methods*, Vol. 7 of *Pharmacochemistry library* (ed. W. Th. Nauta and R. F. Rekker), p. 145. Elsevier, Amsterdam.

Hansch, C. and Leo, A. (1979). *Substituent constants for correlation analysis in chemistry and biology*, p. 58. Wiley, New York.

Hansch, C., Unger, S. H., and Forsythe, A. B. (1973). *Journal of Medicinal Chemistry*, **16**, 1217–22.

Hellberg, S., Sjostrom, M., Skagerberg, B., and Wold, S. (1987). *Journal of Medicinal Chemistry*, **30**, 1126–35.

Herrmann, E. C. (1983). In *Quantitative approaches to drug design*, (ed. J. C. Dearden), pp. 231–2. Elsevier, Amsterdam.

Morgan, E. (1991). *Chemometrics: experimental design.* Wiley, Chichester.

Pleiss, M. A. and Unger, S. H. (1990). In *Quantitative drug design*, (ed. C. A. Ramsden), Vol. 4 of *Comprehensive medicinal chemistry. The rational design, mechanistic study and therapeutic application of chemical compounds*, (ed. C. Hansch, P. G. Sammes, and J. B. Taylor), pp. 561–87. Pergamon Press, Oxford.

Reibnegger, G., Werner-Felmayer, G., and Wachter, H. (1993). *Journal of Molecular Graphics*, **11**, 129–33.

Schaper, K-J. (1983). *Quantitative Structure–Activity Relationships*, **2**, 111–20.

Streich, W. J., Dove, S., and Franke, R. (1980). *Journal of Medicinal Chemistry*, **23**, 1452–6.

Topliss, J. G. and Martin, Y. C. (1975). In *Drug design*, (ed. E. J. Ariens), Vol. 5, pp. 1–21. Academic Press, London.

Unger, S. H. (1980). In *Drug design*, (ed. E. J. Ariens), Vol. 9, pp. 47–119. Academic Press, London.

Wootton, R., Cranfield, R., Sheppey, G. C., and Goodford, P. J. (1975). *Journal of Medicinal Chemistry*, **18**, 607–13.

3
Data pre-treatment

3.1 Introduction

One of the most frequently overlooked aspects of data analysis is consideration of the data that is going to be analysed. How accurate is it? How complete is it? How representative is it? These are some of the questions that should be asked about any set of data, preferably *before* starting to try and understand it, along with the general question 'what do the numbers, or symbols, or categories mean?'

The next few sections discuss some of the more important aspects of the nature and properties of data. It is often the data itself that dictates which particular analytical method may be used to examine it and how successful the outcome of that examination will be.

3.2 The nature of data

So far, in this book the terms descriptor, parameter, and property have been used interchangeably. This can perhaps be justified in that it helps to avoid repetition, but they do actually mean different things and so it would be best to define them here. Physicochemical property refers to a feature of a molecule which is determined by its physical or chemical properties, or a combination of both. Descriptor refers to any means by which a molecule is described or characterized: the term aromatic, for example, is a descriptor, as are the quantities molecular weight and boiling point. Parameter is a term which is used to refer to some numerical measure of a physicochemical property. The two descriptors molecular weight and boiling point are also both parameters; the term aromatic is a descriptor but not a parameter, whereas the question 'how many aromatic rings?' gives rise to a parameter. All parameters are thus descriptors but not vice versa.

In the examples of descriptors and parameters given here, and in previous chapters, it may have been noticed that there are differences in the 'nature' of the values used to express them. This is due to differences in their scales of measurement. It is necessary to consider four different scales of measurement: nominal, ordinal, interval, and ratio. It is important to be aware of the properties of these scales since the nature of the scales determines which analytical methods can be used to treat the data.

Nominal

This is the weakest level of measurement, i.e., has the lowest information content, and applies to the situation where a number or other symbol is used to assign membership to a class. The terms male and female, young and old, aromatic and non-aromatic are all descriptors based on nominal scales. These are dichotomous descriptors, in that the objects (people or compounds) belong to one class or another, but this is not the only type of nominal descriptor. Colour, subdivided into as many classes as desired, is a nominal descriptor as is the question 'which of the four halogens does the compound contain?'.

Ordinal

Like the nominal scale, the ordinal scale of measurement places objects in different classes but here the classes bear some relation to one another, expressed by the term greater than (>). Thus, from the previous example, old > middle-aged > young. Two examples in the context of QSAR are toxic > slightly toxic > non-toxic, and fully saturated > partially saturated > unsaturated. The latter descriptor might also be represented by the number of double bonds present in the structures although this is not chemically equivalent since triple bonds are ignored. It is important to be aware of the situations in which a parameter might appear to be measured on an interval or ratio scale (see below), but because of the distribution of compounds in the set under study, these effectively become nominal or ordinal descriptors (see Section 3.3).

Interval

An interval scale has the characteristics of a nominal scale, but in addition the distances between any two numbers on the scale are of known size. The zero point and the units of measurement of an interval scale are arbitrary: a good example of an interval scale parameter is boiling point. This could be measured on either the Fahrenheit or Celsius temperature scales but the information content of the boiling point values is the same.

Ratio

A ratio scale is an interval scale which has a true zero point as its origin. Mass is an example of a parameter measured on a ratio scale, as are parameters which describe dimensions such as length, volume, etc.

What is the significance of these different scales of measurement? As was mentioned in Section 1.5, many of the well-known statistical methods are parametric, that is, they rely on assumptions concerning the distribution of the data. The computation of parametric tests involves arithmetic manipulation such as addition, multiplication, and division, and this should only be carried out on data measured on interval or ratio scales. When these procedures are used on data measured on other scales they introduce distortions into the data and thus cast doubt on any conclusions which may be drawn from the tests. Non-parametric or 'distribution-free' methods, on the other hand, concentrate on an order or ranking of data and thus can be used with ordinal data. Some of the non-parametric techniques are also designed to operate with classified (nominal) data. Since interval and ratio scales of measurement have all the properties of ordinal scales it is possible to use non-parametric methods for data measured on these scales. Thus, the distribution-free techniques are the 'safest' to use since they can be applied to most types of data. If, however, the data does conform to the distributional assumptions of the parametric techniques, these methods may well extract more information from the data.

3.3 Data distribution

As mentioned in Chapter 1, statistics is often concerned with the treatment of a small* number of samples which have been drawn from a much larger population. Each of these samples may be described by one or more variables which have been measured or calculated for that sample. For each variable there exists a population of samples. It is the properties of these populations of variables that allows the assignment of probabilities, for example, the likelihood that the value of a variable will fall into a particular range, and the assessment of significance (i.e. is one number significantly different from another). Probability theory and statistics are, in fact, separate subjects; each may be said to be the inverse of the other, but for the purposes of this discussion they may be regarded as doing the same job.

How are the properties of the population used? Perhaps one of the most familiar concepts in statistics is the frequency distribution. A plot of a frequency distribution is shown in Fig. 3.1, where the ordinate (y-axis) represents the number of occurrences of a particular value of a variable given by the scales of the abscissa (x-axis). If the data is discrete, usually but not necessarily measured on nominal or ordinal scales, then the

* The term small here may represent hundreds or even thousands of samples. This is a small number compared to a population which is often taken to be infinite.

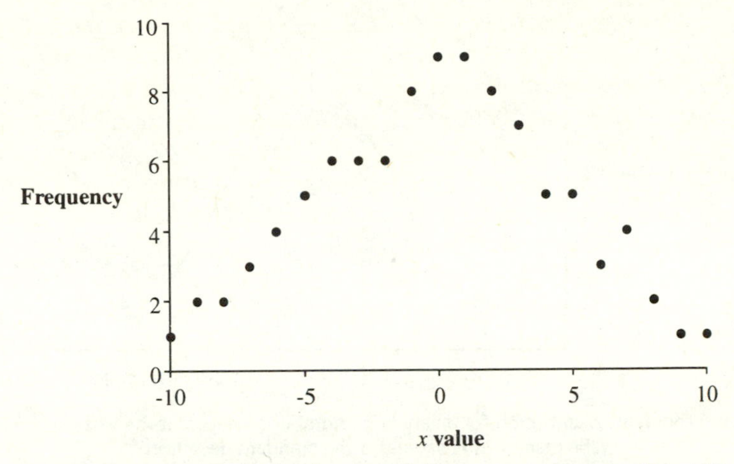

Fig. 3.1. Frequency distribution for the variable x over the range −10 to +10.

variable values can only correspond to the points marked on the scale on the abscissa. If the data is continuous, a problem arises in the creation of a frequency distribution, since every value in the data set may be different and the resultant plot would be a very uninteresting straight line at $y = 1$. This may be overcome by taking ranges of the variable and counting the number of occurrences of values within each range. For the example shown in Fig. 3.2 (where there are a total of 50 values in all), the ranges are 0–1, 1–2, 2–3, and so on up to 9–10.

It can be seen that these points fall on a roughly bell-shaped curve with the largest number of occurrences of the variable occurring around the peak of the curve, corresponding to the mean of the set. If more data

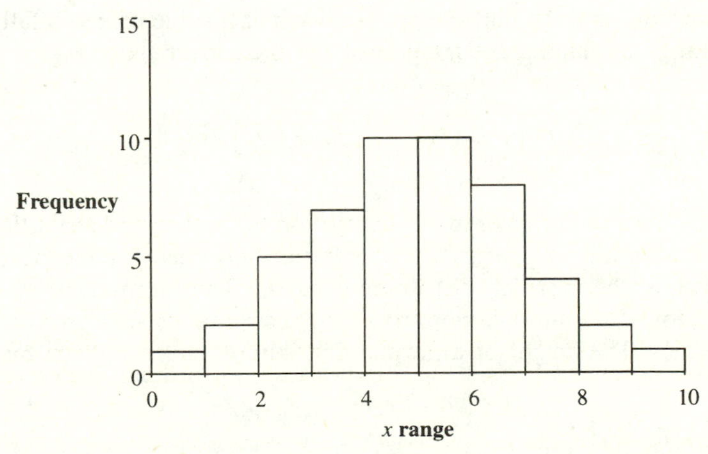

Fig. 3.2. Frequency histogram for the continuous variable x over the range 0 to +10.

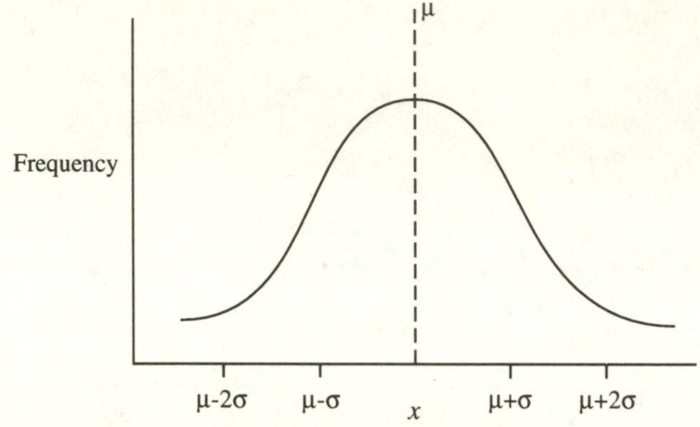

Fig. 3.3. Probability distribution for a very large number of values of the variable x; μ equals the mean of the set and σ the standard deviation.

values are available they will form a smoother curve until in the limit, where a very large number of values are used, we obtain the smooth curve shown in Fig. 3.3.

It is at this point that we see a link between statistics and probability theory. If the height of the curve is standardized so that the area underneath it is unity, the graph is called a probability curve. The height of the curve at some point x can be denoted by $f(x)$ which is called the probability density function (p.d.f.). This function is such that it satisfies the condition that the area under the curve is unity

$$\int_{-\infty}^{\infty} f(x)\,\mathrm{d}x = 1 \tag{3.1}$$

This now allows us to find the probability that a value of x will fall in any given range by finding the integral of the p.d.f. over that range:

$$\mathit{probability}\,(x_1 < x < x_2) = \int_{x_1}^{x_2} f(x)\,\mathrm{d}x \tag{3.2}$$

This brief and rather incomplete description of frequency distributions and their relationship to probability distribution has been for the purpose of introducing the normal distribution curve. The normal or Gaussian distribution is the most important of the distributions that are considered in statistics. The height of a normal distribution curve is given by

$$f(x) = \frac{1}{\sigma\sqrt{2\pi}}\mathrm{e}^{(x-\mu)^2/2\sigma^2} \tag{3.3}$$

This rather complicated function was chosen so that the total area under the curve is equal to 1 for all values of μ and σ. Equation 3.3 has been given so that the connection between probability and the two parameters μ and σ of the distribution can be seen. The curve is shown in Fig. 3.3 where the abscissa is marked in units of σ. It can be seen that the curve is symmetric about μ, the mean, which is a measure of the **location** of the distribution. About one observation in three will lie more than one standard deviation (σ) from the mean and about one observation in 20 will lie more than two standard deviations from the mean. The standard deviation is a measure of the **spread**; it is the two properties, location and spread, of a distribution which allow us to make estimates of likelihood (or 'significance').

Some other features of the normal distribution can be seen by consideration of Fig. 3.4. In part (a) of the figure, the distribution is no longer symmetrical, there are more values of the variable with a higher value. This distribution is said to be skewed, it has a positive skewness; the distribution shown in part (b) is said to be negatively skewed. In part (c) three distributions are overlaid which have differing degrees of 'steepness' of the curve around the mean. The statistical term used to describe the steepness, or degree of peakedness, of a distribution is 'kurtosis'. Various measures may be used to express kurtosis; one known as the **moment ratio** gives a value of three for a normal distribution. Thus it is possible to judge how far a distribution deviates from normality by calculating values of skewness (= 0 for a normal distribution) and kurtosis. As will be seen later, these measures of how well 'behaved' a variable is may be used as an aid to variable selection. Finally, in part (d) of Fig. 3.4 it can be seen that

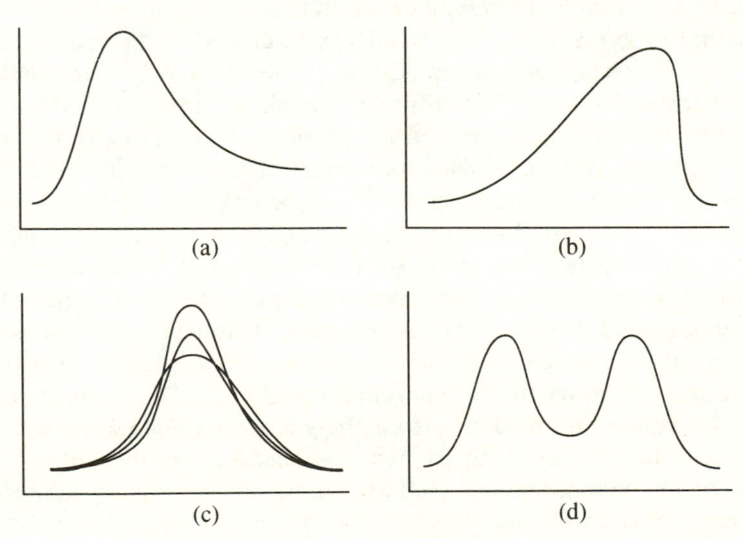

Fig. 3.4. Illustration of deviations of probability distributions from a normal distribution.

the distribution appears to have two means. This is known as a **bimodal** distribution, which has its own particular set of properties distinct to those of the normal distribution.

There are many situations in which a variable that might be expected to have a normal distribution does not. Take for example the molecular weight of a set of assorted painkillers. If the compounds in the set consisted of aspirin and morphine derivatives, then we might see a bimodal distribution with two peaks corresponding to values of around 180 (mol.wt. of aspirin) and 285 (mol.wt. of morphine). Skewed and kurtosed distributions may arise for a variety of reasons, and the effect they will have on an analysis depends on the assumptions employed in the analysis and the degree to which the distributions deviate from normality, or whatever distribution is assumed. This, of course, is not a very satisfactory statement to someone who is asking the question, 'is my data good enough (sufficiently well behaved) to apply such and such a method to it?'. Unfortunately, there is not usually a simple answer to this sort of question. In general, the further the data deviates from the type of distribution that is assumed when a model is fitted, the less reliable will be the conclusions drawn from that model. It is worth pointing out here that real data is unlikely to conform perfectly to a normal distribution, or any other 'standard' distribution for that matter. Checking the distribution is necessary so that we know what type of method can be used to treat the data, i.e., parametric or non-parametric, and how reliable any estimates will be which are based on assumptions of distribution. A caution should be sounded here in that it is easy to become too critical and use a poor or less than 'perfect' distribution as an excuse not to use a particular technique, or to discount the results of an analysis.

Another problem which is frequently encountered in the distribution of data is the presence of outliers. Consider the data shown in Table 3.1 where calculated values of electrophilic superdelocalizability (ESDL10) are given for a set of analogues of antimycin A1, compounds which kill human parasitic worms, *Dipetalonema vitae*. The mean and standard deviation of this variable give no clues as to how well it is distributed and the skewness and kurtosis values of -3.15 and 10.65 respectively might not suggest that it deviates too seriously from normal. A frequency distribution for this variable, however, reveals the presence of a single extreme value (compound 14) as shown in Fig. 3.5. This data was analysed by multiple linear regression (discussed further in Chapter 6), which is a parametric method based on the normal distribution. The presence of this outlier had quite profound effects on the analysis, which could have been avoided if the data distribution had been checked at the outset (particularly by the present author). Outliers can be very informative and should not simply be discarded as so frequently happens. If an outlier is found in one of the descriptor variables (physicochemical data), then it may show

Table 3.1 Physicochemical properties and antifilarial activity of antimycin analogues (from Selwood *et al.* 1990, copyright (1990) American Chemical Society)

Compound number	ESDL10	Calculated log *P*	Melting point (°C)	Activity
1	−0.3896	7.239	81	−0.845
2	−0.4706	5.960	183	−0.380
3	−0.4688	6.994	207	1.398
4	−0.4129	7.372	143	0.319
5	−0.3762	5.730	165	−0.875
6	−0.3280	6.994	192	0.824
7	−0.3649	6.755	256	1.839
8	−0.5404	6.695	199	1.020
9	−0.4499	7.372	151	0.420
10	−0.3473	5.670	195	0.000
11	−0.7942	4.888	212	0.097
12	−0.4057	6.205	246	1.130
13	−0.4094	6.113	208	0.920
14	−1.4855	6.180	159	0.770
15	−0.3427	5.681	178	0.301
16	−0.4597	6.838	222	1.357

that a mistake has been made in the measurement or calculation of that variable for that compound. In the case of properties derived from computational chemistry calculations it may indicate that some basic assumption has been violated or that the particular method employed was not appropriate for that compound. An example of this can be found in semi-empirical molecular orbital methods which are only parameterized for a limited set of the elements. Outliers are not always due to mistakes, however. Consider the calculation of electrostatic potential around a

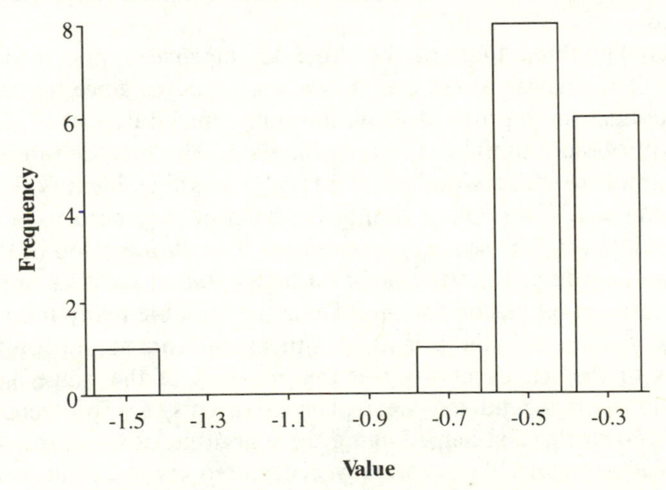

Fig. 3.5. Frequency distribution for the variable ESDL10 given in Table 3.1.

molecule. It is easy to identify regions of high and low values, and these are often used to provide criteria for alignment or as a pictorial explanation of biological properties. The value of an electrostatic potential minimum or maximum, or the value of the potential at a given point, has been used as a parameter to describe sets of molecules. This is fine as long as each molecule in the set has a maximum and/or minimum at approximately the same place. Problems arise if a small number of the structures do not have the corresponding values in which case they will form outliers. The effect of this is to cause the variable, apparently measured on an interval scale, to become a nominal descriptor. Take, for example, the case where 80 per cent of the members of the set have an electrostatic potential minimum of around -50 kcal/mole at a particular position. For the remaining members of the set, the electrostatic potential at this position is zero. This variable has now become an 'indicator' variable which has two distinct values (zero for 20 per cent of the molecules and -50 for the remainder) that identify two different subsets of the data. The problem may be overcome if the magnitude of a minimum or maximum is taken, irrespective of position, although problems may occur with molecules that have multiple minima or maxima. There is also the more difficult philosophical question of what do such values 'mean'.

When outliers occur in the biological or dependent data, they may also indicate mistakes: perhaps the wrong compound was tested, or it did not dissolve, a result was misrecorded, or the test did not work out as expected. However, in dependent data sets, outliers may be even more informative. They may indicate a change in biological mechanism, or perhaps they demonstrate that some important structural feature has been altered or a critical value of a physicochemical property exceeded. Once again, it is best not to simply discard such outliers, they may be very informative.

Is there anything that can be done to improve a poorly distributed variable? The answer is yes, but it is a qualified yes since the use of too many 'tricks' to improve distribution may introduce other distortions which will obscure useful patterns in the data. The first step in improving distribution is to identify outliers and then, if possible, identify the cause(s) of such outliers. If an outlier cannot be 'fixed' it may need to be removed from the data set. The second step involves the consideration of the rest of the values in the set. If a variable has a high value of kurtosis or skewness, is there some good reason for this? Does the variable really measure what we think it does? Are the calculations/measurements sound for all of the members of the set, particularly at the extremes of the range for skewed distributions or around the mean where kurtosis is a problem. Finally, would a transformation help? Taking the logarithm of a variable will often make it behave more like a normally distributed variable, but this is not a justification for always taking logs!

A final point on the matter of data distribution concerns the non-parametric methods. Although these techniques are not based on distributional assumptions, they may still suffer from the effects of 'strange' distributions in the data. The presence of outliers or the effective conversion of interval to ordinal data, as in the electrostatic potential example, may lead to misleading results.

3.4 Scaling

Scaling is a problem familiar to anyone who has ever plotted a graph. In the case of a graph, the axes are scaled so that the information present in each variable may be readily perceived. The same principle applies to the scaling of variables before subjecting them to some form of analysis. The objective of scaling methods is to remove any weighting which is solely due to the units which are used to express a particular variable. An example of this can be seen in the comparison of ^{1}H and ^{13}C NMR shifts. In any comparison of these two types of shifts the variance of the ^{13}C measurements will be far greater simply due to their magnitude. One means by which this can be overcome, to a certain extent at least, is to express all shifts relative to a common structure, the least substituted member of the series, for example. This only partly solves the problem, however, since the magnitude of the Δ shifts will still be greater for ^{13}C than for ^{1}H. A commonly used steric parameter, MR, is often scaled by division by 10 to place it on a similar scale to other parameters such as π and σ.

These somewhat arbitrary scaling methods are far from ideal since, apart from suffering from subjectivity, they require the individual inspection of each variable in detail which can be a time-consuming task. What other forms of scaling are available? One of the most familiar is called **normalization** or **range scaling** where the minimum value of a variable is set to zero and the values of the variable are divided by the range of the variable

$$X'_{ij} = \frac{X_{ij} - X_j(\text{min})}{X_j(\text{max}) - X_j(\text{min})} \tag{3.4}$$

In this equation X'_{ij} is the new range-scaled value for row i (compound i) of variable j. The values of range-scaled variables fall into the range $0 =< X_j =< 1$; the variables are also described as being normalized in the range zero to one. Normalization can be carried out over any preferred range, perhaps for aesthetic reasons, by multiplication of the range-scaled values by a factor. A particular shortcoming of range scaling is that it is dependent on the minimum and maximum values of the variable, thus it is very sensitive to outliers.

Another form of scaling which is less sensitive to outliers is known as **autoscaling** in which the mean is subtracted from the variable values and the resultant values are divided by the standard deviation

$$X'_{ij} = \frac{X_{ij} - \bar{X}_j}{\sigma_j} \tag{3.5}$$

Again, in this equation X'_{ij} represents the new autoscaled value for row i of variable j, $\bar{X}_j$ is the mean of variable j, and σ_j is the standard deviation given by eqn 3.6.

$$\sigma_j = \sqrt{\left(\sum_{i=1}^{N} \frac{(x_{ij} - \bar{x}_j)^2}{N-1}\right)} \tag{3.6}$$

Autoscaled variables have a mean of zero and a variance (standard deviation) of one. Because they are mean centred, they are less susceptible to the effects of compounds with extreme values. That they have a variance of one is useful in variance-related methods (see Chapters 4 and 5) since they each contribute one unit of variance to the overall variance of a data set.

One further method of scaling which may be employed is known as feature weighting where variables are scaled so as to enhance their effects in the analysis. The objective of feature weighting is quite opposite to that of 'equalization' scaling methods described here; it is discussed in detail in Chapter 7.

3.5 Data reduction

This chapter is concerned with the pre-treatment of data and so far we have discussed the nature of data, the properties of the distribution of data, and means by which data may be scaled. All of these matters are important, in so far as they dictate what can be done with data, but perhaps the most important is to answer the question 'what information does the data contain?'. It is most unlikely that any given data set will contain as many pieces of information as it does variables.* That is to say, most data sets suffer from a degree of redundancy and this section describes ways by which redundancy may be identified and, to some extent at least, eliminated. This stage in data analysis is called data reduction in which selected variables are removed from a data set. It should not be confused with dimension

* An example where this is not true is the unusual situation where all of the variables in the set are orthogonal to one another, e.g., principal components (see Chapter 4), but even here some variables may not contain information but be merely 'noise'.

reduction, described in the next chapter, in which high-dimensional data sets are reduced to lower dimensions, usually for the purposes of display.

An obvious first test to apply to the variables in a data set is to look for missing values; is there an entry in each column for every row? What can be done if there are missing values? An easy solution, and often the best one, is to discard the variable but the problem with this approach is that the particular variable concerned may contain information that is useful for the description of the dependent property. An alternative is to provide the missing values, and if these can be calculated with a reasonable degree of certainty, then all is well. If not, however, other methods may be sought. Missing values may be replaced by random numbers, generated to lie in the range of the variable concerned. This allows the information contained in the variable to be used usefully for the members of the set which have 'real' values, but, of course, any correlation or pattern involving that variable does not apply to the other members of the set. An alternative to **random fill** is **mean fill** which, as the name implies, replaces missing values by the mean of the variable involved. This, like random fill, has the advantage that the variable with missing values can now be used; it also has the further advantage that the distribution of the variable will not be altered, other than to increase its kurtosis, perhaps. Another approach to the problem of missing values is to use linear combinations of the *other* variables to produce an estimate for the missing variable. As will be seen later in this section, data sets sometimes suffer from a condition known as multicollinearity in which one variable is correlated with a linear combination of the other variables. This method of filling missing values certainly involves more work, unless the statistics package has it 'built in', and is probably of debatable value since multicollinearity is a condition which is generally best avoided. The ideal solution to missing values is not to have them in the first place!

Another fairly obvious test to apply to the variables in a data set is to identify those parameters which have constant, or nearly constant, values. Such a situation may arise because a property has been poorly chosen in the first place, but may also happen when structural changes in the compounds in the set lead to compensating changes in physicochemical properties. Some data analysis packages have a built-in facility for the identification of such ill-conditioned variables. At this stage in data reduction it is also a good idea to examine the distribution of each of the variables in the set so as to identify outliers or variables which have become 'indicators', as discussed in Section 3.3. Values of the distribution parameters may also be used as decision criteria when choosing which of a pair of correlated variables to retain.

This introduces the correlation matrix. Having removed ill-conditioned variables from the data set, a correlation matrix is constructed by calculation of the correlation coefficient between each pair of variables in the set. A sample correlation matrix is shown in Table 3.2 where the correlation between a pair of variables is found by the intersection of a particular row and column, for example, the correlation between C log *P* and *Iy* is 0.503. The diagonal of the matrix consists of 1s, since this represents the correlation of each variable with itself, and it is usual to show only half of the matrix since it is symmetrical (the top right-hand side of the matrix is identical to the bottom left-hand side). Inspection of the correlation matrix allows the identification of pairs of correlating features, although choice of the level at which correlation becomes important is problematic and dependent to some extent on the requirements of the analysis. There are a number of high correlations ($r > 0.9$) in Table 3.2, however, and removal of one variable from each of these pairs will reduce the size of the data set without much likelihood of removing useful information.

Table 3.2 Correlation matrix for a set of physicochemical properties

	Ix	*Iy*	C log *P*	CMR	CHGE(4)	ESDL(4)	DIPMOM	EHOMO
Ix	1.000							
Iy	0.806	1.000						
C log *P*	0.524	0.503	1.000					
CMR	0.829	0.942	0.591	1.000				
CHGE(4)	0.344	0.349	0.286	0.243	1.000			
ESDL(4)	0.299	0.257	0.128	0.118	0.947	1.000		
DIPMOM	0.337	0.347	0.280	0.233	0.531	0.650	1.000	
EHOMO	0.229	0.172	0.209	0.029	0.895	0.917	0.433	1.000

Having identified pairs of correlated variables, two problems remain in deciding which one of a pair to eliminate. First, is the correlation 'real', in other words has the high correlation coefficient arisen due to a true correlation between the variables or is it caused by some 'point and cluster effect' (see Section 6.2) due to an outlier. The best, and perhaps simplest, way to test the correlation is to plot the two variables against one another, effects due to outliers will then be apparent. It is also worth considering whether the two parameters are likely to be correlated with one another. If one is electronic and the other steric then there is no reason to expect a correlation, although one may exist, of course. On the other hand, maximum width and molecular weight may well be correlated for a set of molecules with similar overall shape.

The second problem, having decided that a correlation is real, concerns the choice of which descriptor to eliminate. One approach to this problem is to delete those features which have the highest number of correlations with other features. This results in a data matrix in which the maximum number of parameters has been retained but in which the inter-parameter correlations are kept low. Another way in which this can be described is to say that the correlation structure of the data set has been simplified. An alternative approach, where the major aim is to reduce the overall size of a data set, is to retain those features which correlate with a large number of others and to remove the correlating descriptors.

Which of these two approaches is adopted depends not only on the data set concerned but also on any chemical (or biological) knowledge concerning the compounds in question. It may be desirable to retain some particular descriptor or group of descriptors on the basis of mechanistic information or hypothesis. It may also be desirable to retain a descriptor because we have confidence in our ability to predict changes to its value with changes in chemical structure; this is particularly true for some of the more 'esoteric' parameters calculated by computational chemistry techniques. What of the situation where there is a pair of correlated parameters and each is correlated with the same number of other features? Here, the choice can be quite arbitrary but one way in which a decision can be made is to eliminate the descriptor whose distribution deviates most from normal. This is used as the basis for variable choice in a published procedure for parameter deletion called CORCHOP; a flow chart for this routine is shown in Fig. 3.6. Although the methods which will be used to analyse a data set once it has been treated as described here may not depend on distributional assumptions, deviation from normality is a reasonable criterion to apply. Interestingly, some techniques of data analysis such as PLS (see Chapter 7) depend on the correlation structure in a data set and may appear to work better if the data is not pre-treated to remove correlations. For ease of interpretation, and generally for ease of subsequent handling, it is recommended that at least the very high correlations are removed from a data matrix.

Another source of redundancy in a data set, which may be more difficult to identify, is where a variable is correlated with a linear combination of two or more of the other variables in the set. This situation is known as **multicollinearity** and may be used as a criterion for removing variables from a data set as part of data pre-treatment. It is also desirable to remove multicollinearity from data sets since this can have adverse effects on the results given by some analytical methods, such as regression analysis (Chapter 6). Factor analysis (Chapter 5) is one method which can be used to identify multicollinearity. Finally, a note of caution needs to be sounded concerning the removal of descriptors based on their correlation with other parameters. It is important to know which variables were discarded because of correlations with others and, if possible, it is best to retain the original starting data set. This may seem like contrary advice since the whole of this chapter has dealt with the matter of simplifying data sets and removing redundant information. However, consider the situation where two variables have a correlation coefficient of 0.7. This represents a shared variance of just under 50 per cent, in other words each variable describes

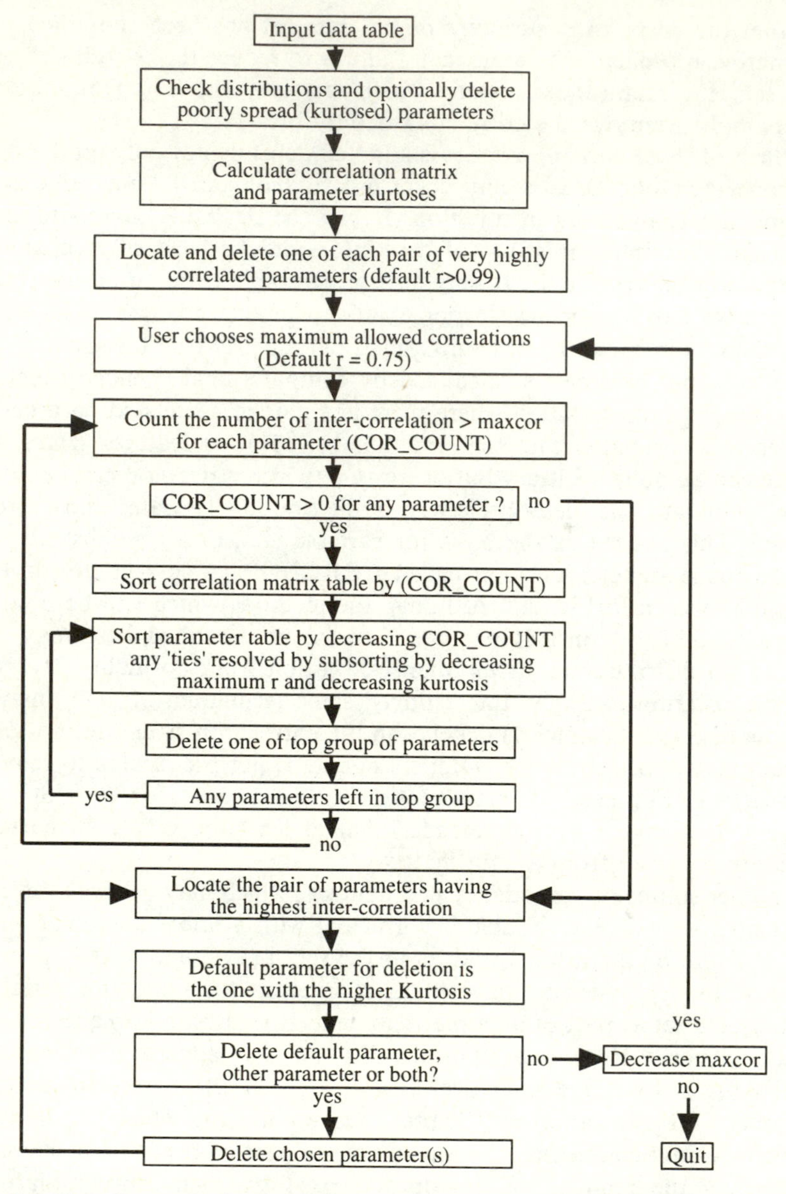

Fig. 3.6. Flow diagram for the correlation reduction procedure CORCHOP (from Livingstone and Rahr 1989, with permission of VCH).

just about half of the information in the other, and this might be a good correlation coefficient cut-off limit for removing variables. Now the correlation coefficient between two parameters also represents the angle between them if they are considered as vectors, as shown in Fig. 3.7. A correlation coefficient of 0.7 is equivalent to an angle of approximately

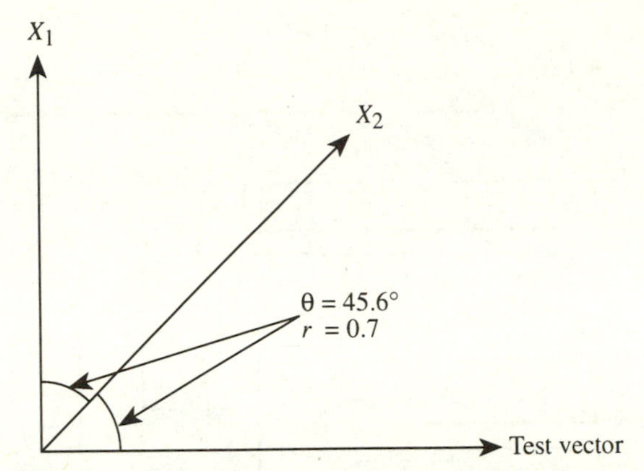

Fig. 3.7. Illustration of the geometric relationship between vectors and correlation coefficients (from Livingstone and Rahr 1989, with permission VCH).

45°. If one of the pair of variables is correlated with a dependent variable with a correlation coefficient of 0.7 this may well be very useful in the description of the property that we are interested in. If the variable that is retained in the data set from that pair is one that correlates with the dependent (X_1 in Fig. 3.7) then all is well. If, however, X_1 was discarded and X_2 retained then this parameter may now be completely uncorrelated ($\theta = 90°$) with the dependent variable. Although this is an idealized case and perhaps unlikely to happen so disastrously in a multivariate data set, it is still a situation to be aware of. One way to approach this problem is to keep a list of all the sets of correlated variables that were in the starting set. Figure 3.8 shows a diagram of the correlations between a set of parameters before and after treatment with the CORCHOP procedure. If no satisfactory correlations with activity are found in the de-correlated set, individual variables can be re-examined using a diagram such as Fig. 3.8. A list of such correlations may also assist when attempts are made to 'explain' correlations in terms of mechanism or chemical features.

3.6 Summary

Selection of the analytical tools which will be used to investigate a set of data should not be dictated by the availability of software on a favourite computer, by what is the current trend, or by personal preference, but rather by the nature of the data within the set. The statistical distribution of the data should also be considered, both when selecting analytical methods to use and when attempting to interpret the results of any analysis. Problems of scaling have been described here and also the almost

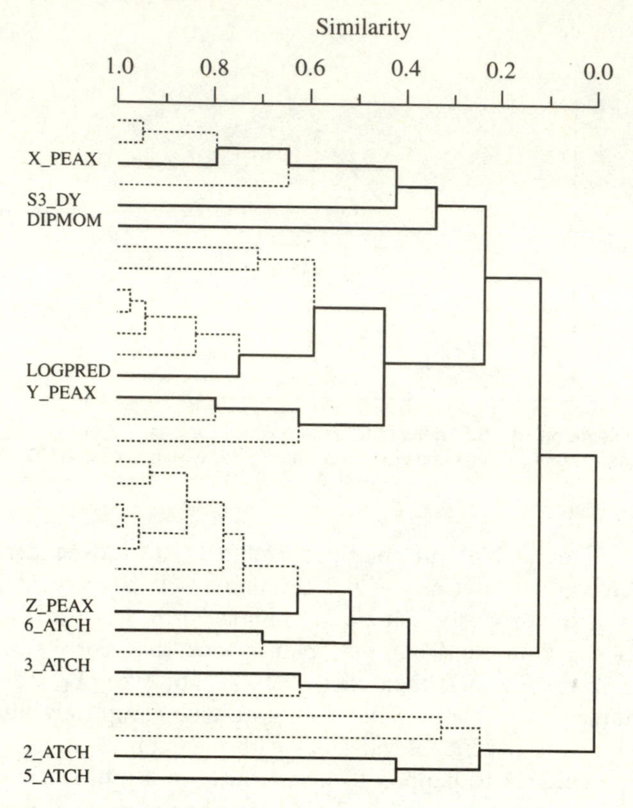

Fig. 3.8. Dendrogram showing the physicochemical descriptors (for a set of antimalarials) retained after use of the CORCHOP procedure. Dotted lines indicate parameters that were present in the starting set (from Livingstone 1989 with permission of the Society of Chemical Industry).

inevitable problem of redundancy, along with some suggestions as to how best to resolve such problems.

References

Livingstone, D. J. (1989). *Pesticide Science*, **27**, 287–304.

Livingstone, D. J. and Rahr, E. (1989). *Quantitative Structure–Activity Relationships*, **8**, 103–8.

Selwood, D. L, Livingstone, D. J., Comley, J. C. W., O'Dowd, A. B., Hudson, A. T., Jackson, P., *et al.* (1990). *Journal of Medicinal Chemistry*, **33**, 136–42.

4
Data display

4.1 Introduction

This chapter is concerned with methods which allow the display of data. The old adage 'a picture is worth a thousand words' is based on our ability to identify visual patterns; it is probably true to say that man is the best pattern-recognition machine that we know of. Unfortunately, we are at our best when operating in only two or three dimensions, although it might be argued that we do operate in higher dimensions if we consider the senses such as taste, smell, touch, and, perhaps, the dimension of time. There are a number of techniques which can help in trying to 'view' multidimensional data and it is perhaps worth pointing out here that this is exactly what the methods do—they allow us to view a data set from a variety of perspectives. If we consider a region of attractive countryside, or a piece of famous architecture such as the Taj Mahal, there is no 'correct' view to take of the scene. There are, however, some views which are 'better' from the point of view of an appreciation of the beauty of the scene, a view of the Taj Mahal which includes the fountains, for example. Figure 4.1 shows a plot of the values of two parameters against one another for a set of compounds which are marked as A for active and I for inactive. Looking at the plot as presented gives a clear separation of the two classes of compounds, the view given by the two parameters is useful. If we consider the data represented by parameter 1, seen from the position marked view 1, it is seen that this also gives a reasonable separation of the two classes although there is some overlap. The data represented by parameter 2 (view 2), on the other hand, gives no separation of the classes at all. This illustrates two important features. First, the consideration of multiple variables often gives a better description of a problem: in this case parameter 2 helps to resolve the conflict in classification given by parameter 1. Second, the choice of viewpoint can be critical and it is usually not possible to say in advance what the 'best' viewpoint will be. Hopefully this simple two-dimensional example has illustrated the problems that may be encountered when viewing multivariate data of 50, 100, or even more dimensions.

Now to multivariate display methods. These methods can conveniently be divided into linear and non-linear techniques as discussed in the

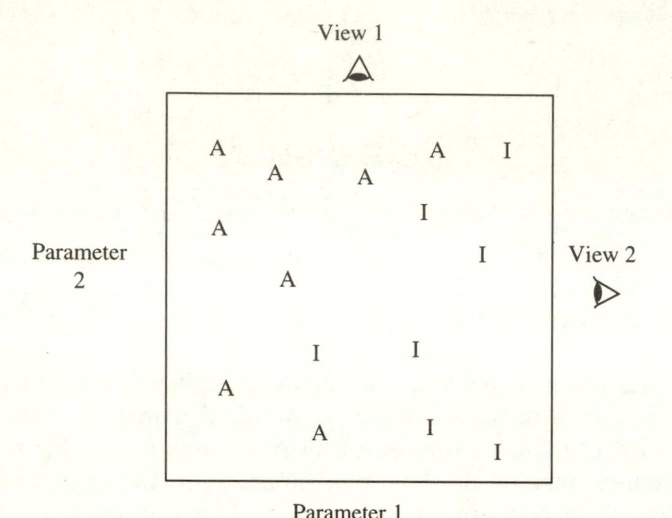

Fig. 4.1. Plot of a set of active (A) and inactive (I) compounds described by two physicochemical properties.

next two sections; cluster analysis as a display method is covered in Chapter 5.

4.2 Linear methods

The simplest and most obvious linear display method is a variable-by-variable plot. The advantages of such plots are that they are very easy to interpret and it is easy to add a new point to the diagram for prediction or comparison (this is not necessarily the case with other methods, as will be shown later). One of the disadvantages of such an approach is that for a multivariate set, there can be many two-dimensional plots, $n(n-1)$ for n variables. Such plots not only take time to generate but also take a lot of time to evaluate. Another disadvantage of this technique is the limited information content of the plots; Fig. 4.1 shows the improvement that can be obtained by the addition of just one parameter to a variable which already describes the biological data reasonably well. How can further dimensions be added to a plot? Computer graphics systems allow the production of three-dimensional pictures which can be viewed in stereo and manipulated in real time. They are often used to display the results of molecular modelling calculations for small molecules and proteins, but can just as easily be adapted to display data. The advantage of being able to manipulate such a display is that a view can be selected which gives the required distribution of data points; in Fig. 4.1, for example, the best view is above the plot. The use of colour or different-shaped symbols can also be used to add extra dimensions to a plot. Figure 4.2 shows a physical model, a reminder of more 'low-tech' times, in which a third parameter is represented by the height of the columns above the baseboard and activity is represented by colour. Another approach is shown in the spectral diagram in Fig. 4.3 which represents a simultaneous display of the activities of compounds (circles) and the relationships between tests (squares); the areas of the symbols represent the mean activity of the compounds and tests. A fuller description of spectral map analysis is given in Chapter 8.

Through these ingenious approaches it is possible to expand diagrams to four, five, or even six dimensions, but this does not even begin to solve the problem of viewing a 50-dimensional data set. What is required is some method to reduce the dimensionality of the data set while retaining its information content. One such technique is known as **principal component analysis (PCA)** and since it forms the basis of a number of useful methods, both supervised and unsupervised, I will attempt to explain it here in some detail. The following description is based, with very grateful permission, on part of Chapter 6 of the book by Hilary Seal (1968).

Fig. 4.2. A physical model used to represent three physicochemical properties, π and σ on the baseboard and MR as the height of the balls. Five colours, indicated by numbers, were used to code the balls (representing compounds) for five activity classes.

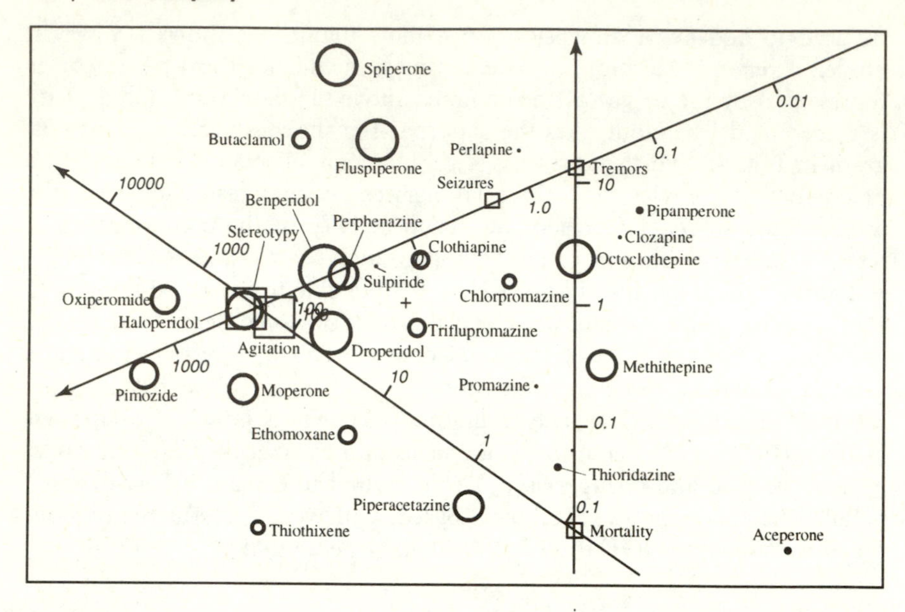

Fig. 4.3. Spectral map of the relationships between the activity of neuroleptics (circles) and *in vivo* tests (squares). (From Lewi 1986 with permission of the *European Journal of Medicinal Chemistry*.)

We have seen in Chapter 3 (Figs 3.1 and 3.2) that a frequency distribution can be constructed for a single variable in which the frequency of occurrence of variable values is plotted against the values themselves. If we take the values of two variables which describe a set of samples (compounds, objects, mixtures, etc.) a frequency distribution can be shown for both variables simultaneously (Fig. 4.4). In this diagram the height of the surface represents the number of occurrences of samples which have variable values corresponding to the X and Y values of the plane which the surface sits on. The highest point of this surface, the summit of the hill, corresponds to the mean of each of the two variables. It is possible to take slices through a solid object such as this and plot these as ellipses on a two-dimensional plot as shown in Fig. 4.5. These ellipses represent population contours: as the slices are taken further down the surface from the summit, they produce larger ellipses which contain higher proportions of the population of variable values. Two important things can be seen from this figure. First, the largest axis of the ellipses corresponds to the variable (X_1) with the larger standard deviation. Thus, the greatest part of the shape of each ellipse is associated with the variable which contains the most variance, in other words, information. Second, the two axes of the ellipses are aligned with the two axes of the plot. This is because the two variables are not associated with one another; where there are high values of variable X_2, there is a spread of values of variable X_1 and vice versa. If

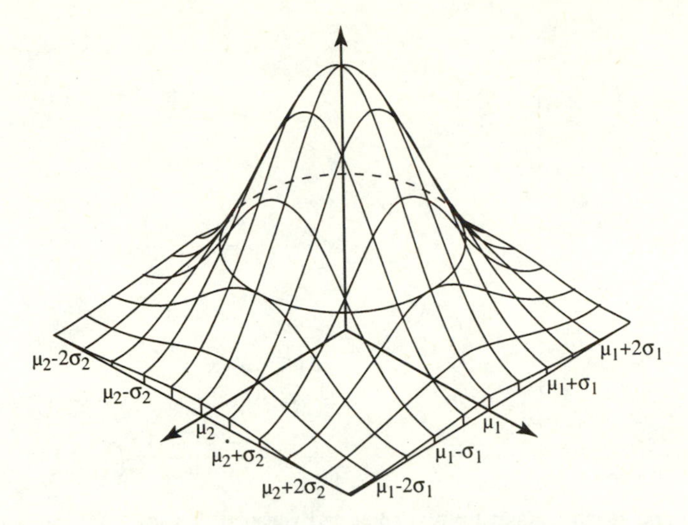

Fig. 4.4. Three-dimensional plot of the frequency distribution for two variables. The two variables have the same standard deviations so the frequency 'surface' is symmetrical. (From Seal 1968 with permission of Methuen.)

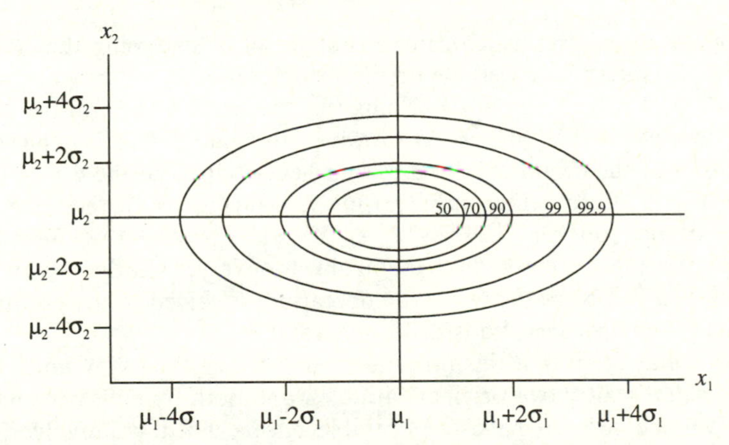

Fig. 4.5. Population contours from a frequency distribution such as that shown in Fig. 4.4. In this case, the variables have different standard deviations ($\sigma_1 > \sigma_2$) so the contours are ellipses (from Seal 1968 with permission of Methuen).

the two variables are correlated then the ellipses are tilted as shown in Fig. 4.6 where one population contour is plotted for two variables, Y and X, which are positively correlated. The two 'ends' of the population ellipse are located in two quadrants of the X–Y space which correspond to (low X, low Y) and (high X, high Y). If the variables were negatively correlated, the ellipse would be tilted so that high values of Y correspond to low values of X and the other end of the ellipse would be in the (low Y, high X) quadrant. The relationship between the two axes, X and Y, and

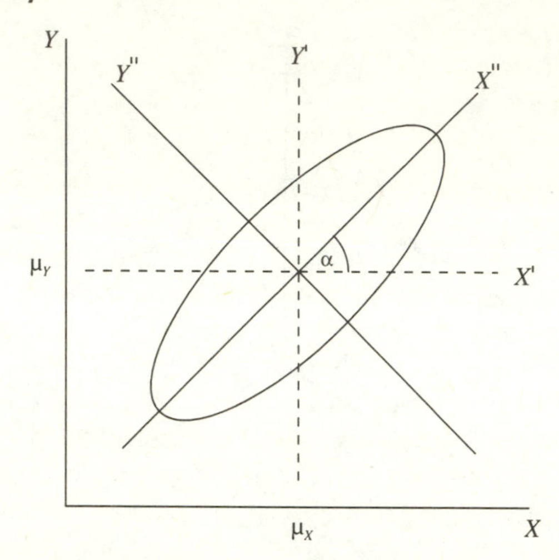

Fig. 4.6. Population contour for two correlated variables, X and Y. The axes X' and Y' represent mean-centred variables achieved by translation of the origin of X and Y. The axes X'' and Y'' are formed by rotation of X' and Y' through the angle α (after Seal 1968 with permission of Methuen and Co.).

the population ellipse, which can be thought of as enclosing the cloud of data points, shows how well the original variables describe the information in the data. Another way of describing the data is to transform the axes X and Y to new axes X' and Y' as shown in the figure. This is achieved by translation of the origin of X and Y to a new position in the centre of the ellipse, a procedure called, unsurprisingly, centring. A further operation can be carried out on the new axes, X' and Y', and that is rotation through the angle α marked on the figure, to give yet another set of axes, X'' and Y''. These are the two basic operations involved in the production of **principal components**, translation and rotation.

Now it may seem that this procedure has not achieved very much other than to slightly alter two original variables, and both by the same amount, but it will be seen to have considerable effects when we involve larger numbers of variables. For the present, though, consider the results of this procedure as it illustrates some important features of PCA. The new variable, X'', is aligned with the major axis of the ellipse and it is thus explaining the major part of the variance in the data set. The other new variable, Y'', is aligned with the next largest axis of the ellipse and is thus explaining the next largest piece of information in the set of data points. Why is this the next largest piece of variance in the data set? Surely another direction can be found in the ellipse which is different to the major axis? The answer to this question is yes, but a requirement of principal components is that they are orthogonal to one another (also uncorrelated) and in this two-dimensional example that means at 90°. The two important properties of principal components are:

(1) the first principal component explains the maximum variance in the data set, with subsequent components describing the maximum part of the remaining variance subject to the condition that
(2) all principal components are orthogonal to one another.

In this simple two-dimensional example it is easy to see the directions that the two principal components (PCs) must take to describe the variance in the data set. Since the two axes, X and Y, were originally orthogonal it is also easy to see that it is only necessary to apply the same single rotation to each axis to produce the PCs. In the situation where there are multiple variables, the same single rotation (including reflection) is applied to all the variables. The other feature of principal component analysis that this example demonstrates is the matter of dimensionality. The maximum number of components which are orthogonal and that can be generated in two dimensions is two. For three dimensions, the maximum number of orthogonal components is three, and so on for higher dimensional data sets. The other 'natural' limit for the number of components that can be extracted from a multidimensional data set is dictated by the number of data points in the set. Each PC must explain some part of the variance in the data set and thus at least one sample point must be associated with each new PC dimension. The third condition for PCA is thus

(3) as many PCs may be extracted as the smaller of p (data points) or n (dimensions) for a $p \times n$ matrix (denoted by q in eqn (4.1)).*

There are other important properties of PCs to consider, such as their physical meaning and their 'significance'. These are discussed further in this section and in Chapter 7; for the present it is sufficient to regard them as means by which the dimensionality of a high-dimensional data space can be reduced. How are they used? In the situation where a data set contains many variables the PCs can be regarded as new variables created by taking a linear combination of the original variables as shown in eqn (4.1).

$$\begin{aligned} PC_1 &= a_{1,1}v_1 + a_{1,2}v_2 + \ldots a_{1,n}v_n \\ PC_2 &= a_{2,1}v_1 + a_{2,2}v_2 + \ldots a_{2,n}v_n \\ \\ PC_q &= a_{q,1}v_1 + a_{q,2}v_2 + \ldots a_{q,n}v_n \end{aligned} \tag{4.1}$$

Where the subscripted term, a_{ij}, represents the contribution of each original variable ($v_1 \rightarrow v_n$) in the N-dimensional set to the particular principal component ($1 \rightarrow q$) where q (the number of principal components) is the smaller of the p points or n dimensions. These coefficients have a sign

* Actually, it is the rank of the matrix, denoted by $r(\mathbf{A})$, which is the maximum number of linearly independent rows (or columns) in $\mathbf{A}$. $0 \leq r(\mathbf{A}) \leq \min(p, n)$, where $\mathbf{A}$ has p rows and n columns.

associated with them, indicating whether a particular variable makes a negative or positive contribution to the component, and their magnitude shows the degree to which they contribute to the component. The coefficients are also referred to as loadings and represent the contribution of individual variables to the principal components.* Since the PCs are new variables it is possible to calculate values for each of these components for each of the objects (data points) in the data set to produce a new (reconstructed but related) data set. The numbers in this data set are known as principal component scores; the process is shown diagrammatically in Fig. 4.7. Now, it may not seem that this has achieved much in the way of dimension reduction: while it is true that the scores matrix has a 'width' of q this will only be a reduction if there were fewer compounds than variables in the starting data set. The utility of PCA for dimension reduction lies in the fact that the PCs are generated so that they explain maximal amounts of variance. The majority of the information in many data sets will be contained in the first few PCs derived from the set. In fact, by definition, the most informative view of a data set, in terms of variance at least, will be given by consideration of the first two PCs. Since the scores matrix contains a value for each compound corresponding to each PC it is possible to plot these values against one another to produce a low-dimensional picture of a high-dimensional data set. Figure 4.8 shows a scores plot for 13 compounds described by 33 calculated physicochemical

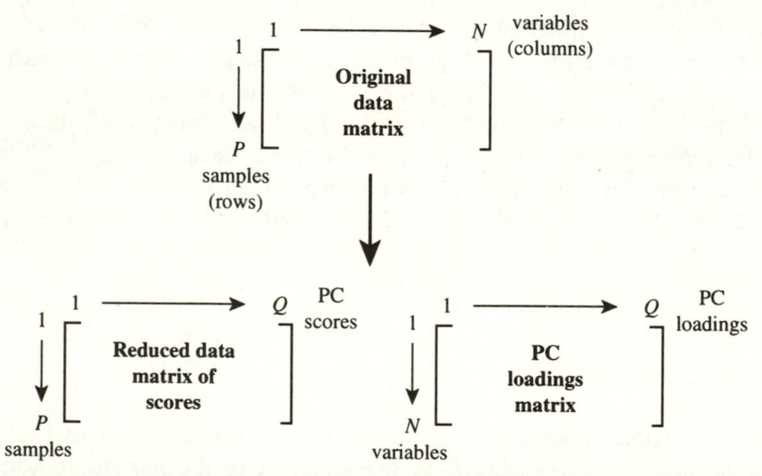

Fig. 4.7. Illustration of the process of principal components analysis to produce a 'new' data matrix of Q scores for P samples where Q is equal to (or less than) the smaller of N (variables) or P (samples). The loadings matrix contains the contribution (loading) of each of the N variables to each of the Q principal components.

---

* A loading is actually the product of the coefficient and the eigenvalue of the principal component (a measure of its importance) as described later.

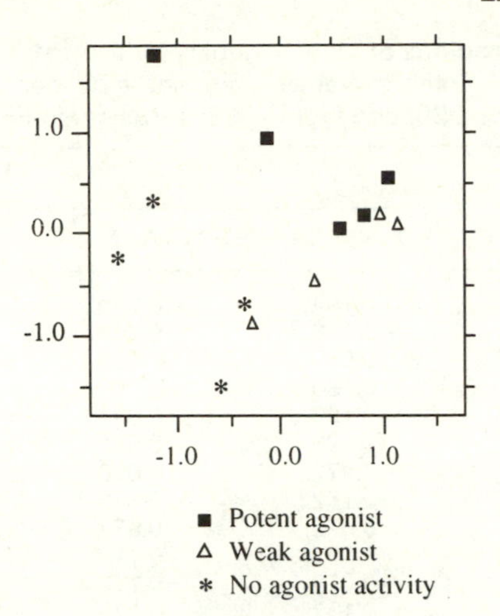

Fig. 4.8. Scores plot for 13 analogues of γ-aminobutyric acid (from Hudson *et al.* 1989, copyright (1989) Wellcome Foundation).

properties. This picture is drawn from the scores for the first two PCs and it is interesting to see that the compounds are roughly separated into three classes of biological activity—potent, weak, and no agonist activity. Although the separation between classes is not ideal this is still quite an impressive picture since it is an example of unsupervised learning pattern recognition; the biological information was not used in the generation of the PCs. Table 4.1 gives a list of the loadings of the original 33 variables with the first three PCs. This table should give some idea of the complex nature of PCs derived from large dimensional data sets. Some variables contribute in a negative fashion to the first two PCs, e.g., CMR, 4-ESDL, 3-NSDL, and so on, while others have opposite signs for their loadings on these two PCs. The change in sign for the loadings of an individual variable on two PCs perhaps seems reasonable when we consider that the PCs are orthogonal; the PCs are taking different 'directions' and thus a variable that contributes positively to one PC might be negatively associated with another (see Fig. 4.13). Where the signs of the loadings of one variable on two PCs are the same, the loading for that variable on a third PC is often (but not always) reversed, demonstrating that the third component is taking a different direction to the first two. It should be pointed out here that the direction that a PC takes, with respect to the original variables, is arbitary. Reversing all of the signs of the loadings of the variables on a particular PC produces a component which explains the same amount of variance. When PCs are calculated for the same data set using two different software packages, it is not unusual to find that the signs of the loadings of the variables on corresponding PCs (e.g., the first PC from the two programs) are reversed, but the eigenvalues (variance

Table 4.1 Loadings of input variables for the first three principal components (total explained variance = 70 per cent) (from Hudson *et al.* 1989, copyright (1989) Wellcome Foundation)

	Loading		
Variable	PC_1	PC_2	PC_3
CMR	−0.154	−0.275	0.107
1-ATCH	−0.196	0.096	0.298
2-ATCH	0.015	−0.203	−0.348
3-ATCH	0.186	0.003	0.215
4-ATCH	−0.183	0.081	−0.197
5-ATCH	−0.223	0.061	−0.027
6-ATCH	0.272	−0.030	0.049
X-DIPV	0.152	−0.085	−0.059
Y-DIPV	0.079	−0.077	−0.278
Z-DIPV	0.073	−0.117	0.019
DIPMOM	−0.019	0.173	−0.006
T-ENER	0.137	0.146	−0.242
1-ESDL	0.253	−0.156	0.120
2-ESDL	0.221	−0.071	−0.020
3-ESDL	−0.217	0.108	−0.248
4-ESDL	−0.167	−0.115	−0.245
5-ESDL	−0.105	0.197	0.158
6-ESDL	0.128	0.072	0.337
1-NSDL	0.183	−0.253	0.148
2-NSDL	0.186	−0.236	0.025
3-NSDL	−0.021	−0.136	−0.365
4-NSDL	−0.226	0.195	−0.046
5-NSDL	−0.111	0.227	0.141
6-NSDL	−0.099	0.257	0.125
VDW_VOL	−0.228	−0.229	0.031
X-MOFI	−0.186	−0.238	0.136
Y-MOFI	−0.186	−0.266	0.093
Z-MOFI	−0.209	−0.238	0.090
X-PEAX	−0.218	−0.178	−0.020
Y-PEAX	−0.266	−0.050	0.084
Z-PEAX	−0.051	−0.217	0.035
MOL_WT	−0.126	−0.263	0.189
IHET_1	0.185	−0.071	−0.052

explained) are the same. The other important piece of information to note in Table 4.1 is the magnitude of the coefficients. Many of the variables that make a large contribution to the first component will tend to have a small coefficient in the second component and vice versa. Some variables, of course, can make a large contribution to both of these PCs, e.g., CMR, T-ENER, 1-ESDL, 1-NSDL, etc., in which case they are likely to make a smaller contribution to the third component. The variable T-ENER demonstrates an exception to this in that it has relatively high loadings on all three components listed in the table.*

* For this data set it is possible to calculate a total of 13 components although not all are 'significant' as discussed later.

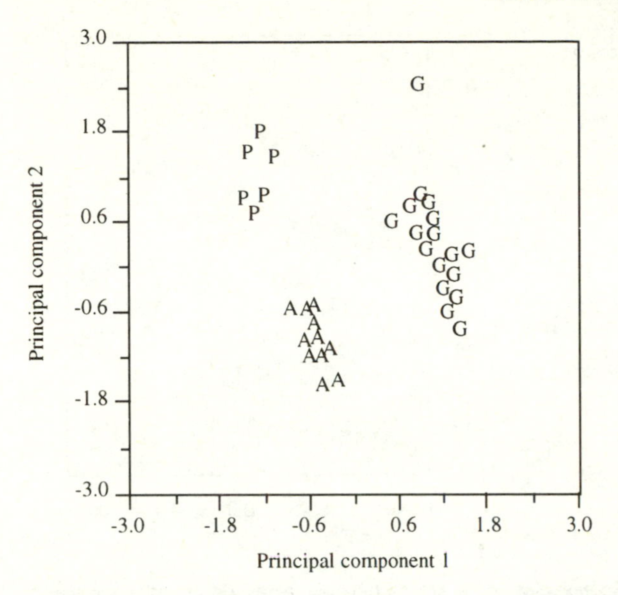

Fig. 4.9. Scores plot for fruit juice samples: A, apple juice; P, pineapple juice; and G, grape juice (from Dizy *et al.* 1992, with permission of the Society of Chemical Industry).

Figure 4.9 shows an example of the value of PCA in food science. This is a scores plot for the first two PCs derived from a data set of 15 variables measured on 34 samples of fruit juices. The variables included pH, total phenols, reducing sugars, total nitrogen, ash, glucose content, and formol number, and the samples comprised 17 grape, 11 apple, and six pineapple juices. As can be seen from the figure, the first two components give a very satisfactory separation of the three types of juice. The first PC was related to richness in sugar since the variables reducing sugars, total sugars, glucose, °Brix, dry extract, and fructose load highly onto it. This component distinguishes grape from apple and pineapple juice. The second PC, which separates apple from pineapple juice was highly correlated with the glucose : fructose ratio, total nitrogen, and formol number. In this example it is possible to attempt to ascribe some chemical 'meaning' to a PC, here sugar richness described by PC_1, but in general it should be borne in mind that PCs are mathematical constructs without necessarily having any physical significance. An example of the use of PCA in another area of chemical research is shown in Fig. 4.10. This scores plot was derived from PCA applied to a set of seven parameters, calculated log *P* and six theoretical descriptors, used to describe a series of 14 substituted benzoic acids. The major route of metabolism of these compounds in the rat was determined by NMR measurements of urine, or taken from the literature, and they were assigned to glycine (Fig. 4.11a) or glucuronide (Fig. 4.11b) conjugates. A training set of 12 compounds is shown on the scores plot in

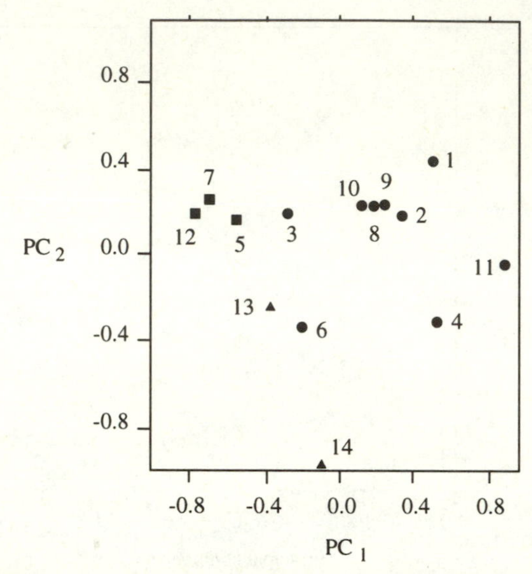

Fig. 4.10. Scores plot for a set of benzoic acids described by seven physicochemical properties. Compounds are metabolized by the formation of glucuronide conjugates (squares) or glycine conjugates (circles) and two test set compounds are shown (triangles). (From Ghauri *et al.* 1992, copyright (1992) Wellcome Foundation.)

Fig. 4.10 where it can be seen that the glucuronide conjugate-forming compounds (squares) are well separated from the rest of the set. Two test set compounds are shown as triangles; compound 13 is metabolized by the glucuronide route and does lie close to the other glucuronide conjugate formers. However, this compound is also close, in fact closer, to a glycine conjugate-forming acid (number 6) and thus might be predicted to be metabolized by this route. The other test set compound lies in a region of PC space, low values of PC_2, which is not mapped by the other compounds in the training set. This compound is metabolized by the glycine conjugate route but it is clear that this could not be predicted from this

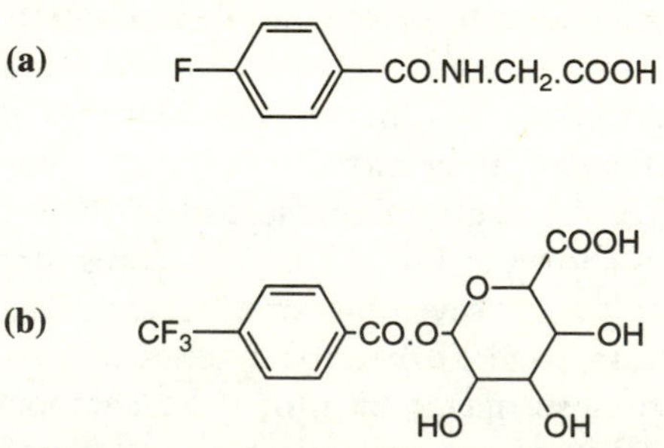

(a) F—C_6H_4—CO.NH.CH_2.COOH

(b) CF_3—C_6H_4—CO.O— (glucuronic acid ring: O, COOH, OH, OH, HO)

Fig. 4.11. Structure of (a) the glycine conjugate of 4-fluorobenzoic acid; (b) the glucuronic acid conjugate of 4-trifluoromethylbenzoic acid. (From Ghauri *et al.* 1992, copyright (1992) Wellcome Foundation.)

scores plot. This example serves to illustrate two points. First, the PC scores plot can be used to classify successfully the metabolic route for the majority of these simple benzoic acid analogues, but that individual predictions may not be unambiguous. Second, it demonstrates the importance of careful choice of test and training set compounds. Compound 14 must have some extreme values in the original data matrix and thus might be better treated as a member of the training set. In fairness to the original report it should be pointed out that the selection of 'better' variables, in terms of their ability to classify the compounds, led to plots with much better predictive ability.

As was seen in Table 4.1, PCA not only provides information about the relationships between samples in a data set but also gives us insight into the relationships between variables. The schematic representation of PCA in Fig. 4.7 shows that the process produces two new matrices, each of width Q, where Q is the smaller of N (variables) or P (samples). The scores matrix contains the values of new variables (scores) which describe the samples. The loadings matrix contains the values of the loadings (correlations) of each variable with each of the Q principal components. These loadings are the coefficients for the variables in eqn (4.1) and can be used to construct a loadings plot for a pair of PCs. In an analysis of an extensive set of physicochemical substituent constants, Van de Waterbeemd and colleagues (1989) produced the PC loadings shown in Table 4.2. The loadings for the full set of substituent constants are shown projected onto the first two PC axes in Fig. 4.12. In this figure each point represents a variable; where points are clustered together, the variables are all highly associated with one another. Those points which lie close to the origin of the plot (e.g. 38, 43, 44, and 50) make little contribution to either PC,

Table 4.2 Loadings for selected variables on the first two PCs* (reproduced with kind permission from Van de Waterbeemd *et al.* 1989, copyright (1989) ESCOM Science Publishers B.V.)

No. on Fig. 4.12	Parameter	PC_1	PC_2
1	PIAR	−0.72	0.11
2	PIAL	−0.66	0.16
3	FARR	−0.71	0.23
4	FALR	−0.69	0.21
5	FARHL	−0.72	0.17
6	FALHL	0.69	0.23
7	K	0.04	−0.49
38	SX	0.14	0.04
43	RE	0.07	0.01
44	I	0.14	0.07
50	HD	0.12	0.07
59	RAND	0.09	0.39

* From a total of 75 parameters describing 59 substituents.

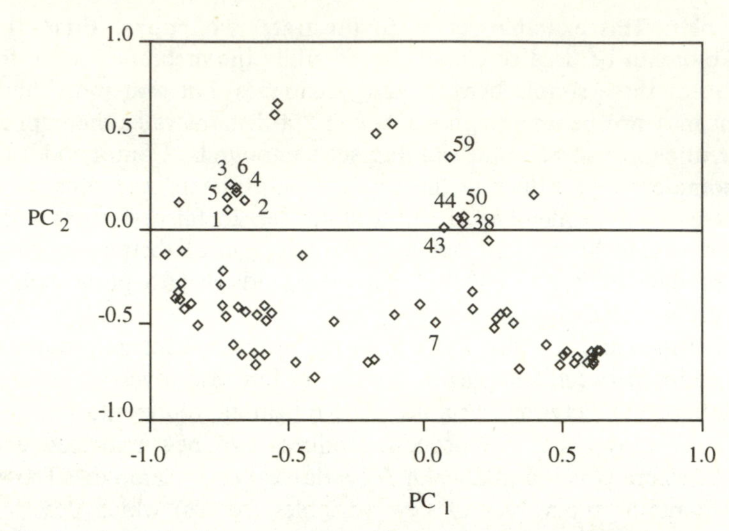

Fig. 4.12. Loadings plot for a set of 75 substituent constants (reproduced with kind permission from Van de Waterbeemd *et al.* 1989, copyright (1989) ESCOM Science Publishers B.V.).

conversely the points at the extremes of the four quadrants are highly associated with their respective PCs. The cluster of variables represented by points 1 to 6 is a chemically 'sensible' one, these are all descriptors of lipophilicity. The fact that parameter 59 lies close to the origin is reassuring, this variable was generated from random numbers. Descriptor 7 is derived from measurements of charge-transfer complexes, its relationship to other parameters is examined further in Section 7.3. Points which lie on their own in the PC space represent variables which contain some unique information not associated with other variables.

By joining the points representing variables to the origin of the PC plot it is possible to construct vectors in the two-dimensional plane of PC space. This type of representation can be adapted to produce a diagram which aims to give another, more visual, explanation of principal component analysis. In Fig. 4.13 the solid arrows represent individual variables as vectors, with the length of each arrow proportional to the variance contained in the variable. This is not the same type of plot as Fig. 4.12; the two-dimensional space is not PC space but is intended to represent N dimensions. The position of the arrows in the diagram demonstrates the relationships between the variables, arrows which lie close to one another represent correlated variables. The first PC is shown as a dotted arrow and it can be seen to lie within a cluster of correlated variables. The loadings of these variables (and the others in the set) are found by projection of the arrows onto this PC arrow, illustrated for just two variables for clarity. The length of the PC arrow is given by vector addition of the arrows

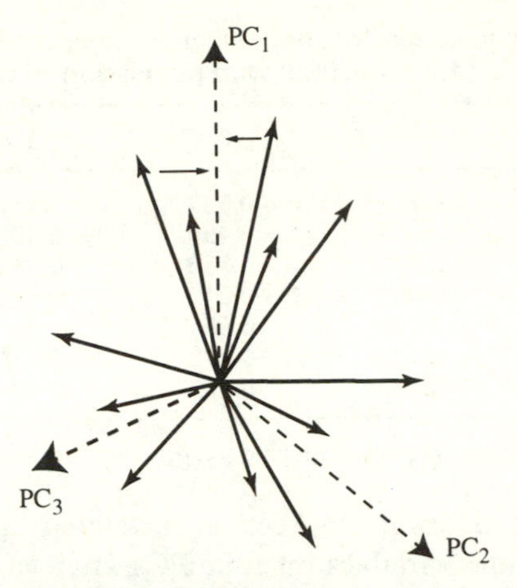

Fig. 4.13. Pictorial representation of the relationship between data vectors (variables), shown by solid lines, and PCs shown by dotted lines. The plane of the diagram is not 'real' two-dimensional space or PC space but is meant to represent *N* dimensions.

representing the variables and, as for the individual variables, this represents the variance contained in this component. The second and third PCs lie within other sets of correlated variables and are shorter vectors than the first since they are explaining smaller amounts of variance in the set. The PC vectors are not at right angles (orthogonal to one another) in this diagram since the space in the figure is not 'real' two-dimensional space. The relationship between PC vectors and the variable vectors illustrates an operation that can be carried out on PCs in order to simplify their structure. This can be of assistance in attempts to interpret PCs and may also result in PCs which are better able to explain some dependent variable. The three PC vectors shown in Fig. 4.13 were generated so as to explain the maximum variance in the data set and thus there are a lot of variables associated with them. This association of many variables with each component leads to low loadings for some of the variables, particularly some of the more 'important' (high-variance) variables. By trying to explain the maximum amount of variance in the set, PCA achieves a compromise between PC 'directions' that are aligned with high-variance variables and directions that are aligned with a large number of variables. Rotation of the PCs allows new directions to be found in which fewer variables are more highly associated with each PC. There are a number of techniques available to achieve such rotations, one of the commonest is known as *varimax* rotation (Jackson 1991). Table 4.3 shows the loadings of seven physicochemical parameters on four PCs for a set of 18 naphthalene derivatives. High loadings, i.e., variables making a large contribution, for each component are shown in bold type. It can be seen that the first component in particular has a quite complicated structure with four variables contributing to it and that two of these, π and MR, are

Table 4.3 Parameter loadings for four principal components (from Livingstone 1991, after Schultz and Moulton 1985, with permission of Springer-Verlag)

Parameter	1	2	3	4
π	**0.698***	−0.537	−0.121	−0.258
MR	**0.771**	0.490	−0.302	−0.002
F	0.261	0.389	**0.745**	−0.423
R	0.405	−0.012	0.578	**0.697**
H_a	−0.140	**0.951**	0.071	−0.101
H_d	**−0.733**	0.373	−0.271	0.172
${}^1\chi^v_{sub}$	**0.739**	0.412	−0.404	0.163

* Boldface numbers indicate parameters making a large contribution to each component.

properties that it is desirable to keep uncorrelated. Table 4.4 shows the loadings of these same variables on four PCs after varimax rotation. The structure of the first PC has been simplified considerably and the correlation between π and MR has been almost eliminated by reducing the π loading from 0.6988 to 0.2. This parameter now loads onto the second PC (note the change in sign) and the properties which were highly associated with the third and fourth PCs have had their loadings increased. Varimax rotation results in a new set of components, often referred to as factors, in which loadings are increased or reduced to give a simplified correlation structure. This rotation is orthogonal, that is to say the resulting factors are orthogonal like the PCs they were derived from. Other orthogonal rotations may be used to aid in the interpretation of PCs and non-orthogonal (oblique) rotations also exist (Jackson 1991).

Table 4.4 Parameter loadings after varimax rotation (from Livingstone 1991, after Schultz and Moulton 1985, with permission of Springer-Verlag)

Parameter	1	2	3	4
π	0.200	**0.919***	0.012	0.012
MR	**0.891**	0.195	0.093	0.061
F	0.020	−0.003	**0.975**	0.123
R	0.081	0.018	0.115	**0.982**
H_a	0.272	0.451	0.318	−0.083
H_d	−0.159	−0.285	−0.138	−0.148
${}^1\chi^v_{sub}$	**0.974**	0.037	−0.024	0.064

* Boldface numbers indicate parameters making a large contribution to each component.

Scores or loadings plots are not restricted to the first two PCs, although all of the examples shown so far have been based on the first two PCs. By definition, the first two PCs explain the largest amount of variance in a data set, but plots of other components may be more informative; Section 7.3.1, for example, shows a data set where the first and fourth PCs were most useful in the explanation of a dependent variable. Plots are also not restricted to just two PCs, although two-dimensional plots are quite popular since they fit easily onto two-dimensional paper! The physical model shown earlier (Fig. 4.2) is a four-dimensional plot and the spectral map (Fig. 4.3) contains a third dimension in the thickness of the symbols. Figure 4.14 shows a plot of the first three PCs calculated from a GC–MS analysis (32 peaks) of natural orange aroma samples. The different samples, labelled A to P, were of distinct types of orange aroma provided by six different commercial flavour houses. These orange aromas could be classified into nine separate categories, as indicated by the different symbols on the plot, and it can be seen that this three-dimensional diagram separates the categories quite well.

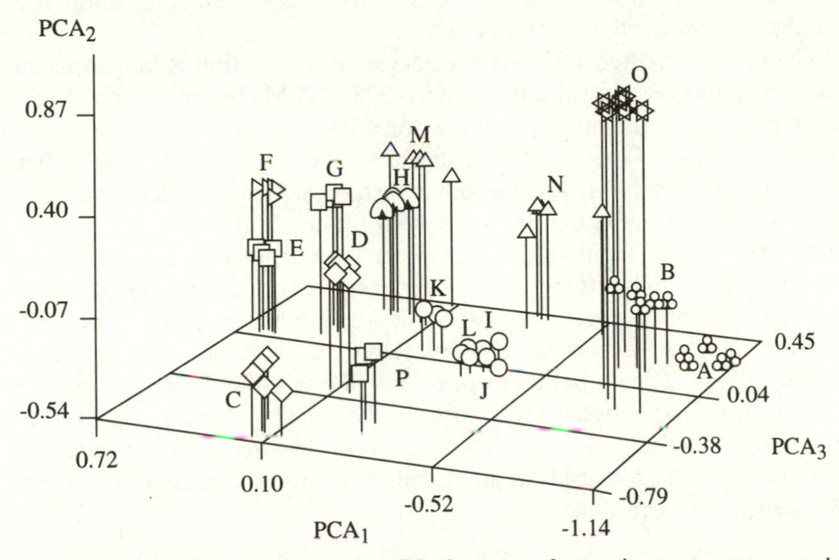

Fig. 4.14. Scores plot on the first three PCs for a set of natural orange aroma samples described by GC–MS data. Different samples are indicated by the letters A–P, and different categories by different symbols (from Lin *et al.* 1993 with permission of Elsevier Science).

As mentioned at the beginning of this section, PCA lies at the heart of several analytical methods which will be discussed in later chapters. Some other features of PCs, such as their 'significance', are also discussed later in the book; this section has been intended to illustrate the use of PCA as a linear dimension reduction method.

4.3 Non-linear methods

For any given data set of points in N dimensions it is possible to calculate the distances between pairs of points by means of an equation such as that shown in eqn (4.2).

$$d_{ij} = \sqrt{\left(\sum_{k=1,N} (d_{i,k} - d_{j,k})^2\right)} \tag{4.2}$$

This is the expression for the Euclidean distance where d_{ij} refers to the distance between points i and j in an N-dimensional space given by the summation of the differences of their coordinates in each dimension ($k = 1, N$). Different measures of distance may be used to characterize the similarities between points in space, e.g., city-block distances, Mahalonobis distance (see Digby and Kempton 1987 for examples), but for most purposes the familiar Euclidean distance is sufficient. The collection of interpoint distances is known, unsurprisingly, as a distance matrix (see Section 5.2 for an example) and this is used as the starting point for a number of multivariate techniques.

The display method which is considered in this section is known as **non-linear mapping** (Kowalski and Bender 1973), NLM for short, and takes as its starting point the distance matrix for a data set calculated according to eqn (4.2). The distances in this distance matrix are labelled d^*_{ij} to indicate that they relate to N-space interpoint distances. Having calculated the N-space distance matrix, the next step is to randomly (usually, but see later) assign the points (compounds, samples) to positions in a lower dimensional space. This is usually a two-dimensional space for ease of plotting but can be a three-dimensional space if a computer graphics display with stereo is used, or a two-dimensional stereo plot with appropriate viewer. Having assigned the p points to positions in a two-dimensional coordinate system, distances between the points can be calculated using eqn (4.2) and these are labelled d_{ij}. The difference between the N-space interpoint distances and the 2-space interpoint distances can be expressed as an error, E, as shown in eqn (4.3).

$$E = \sum_{i>j} (d^*_{ij} - d_{ij})^2 / (d^*_{ij})^\rho \tag{4.3}$$

Minimization of this error function results in a two-dimensional display of the data set in which the distances between points are such that they best represent the distances between points in N-space. The significance of the power term, ρ, will be discussed later in this section: it serves to alter the emphasis on the relative importance of large versus small N-space interpoint distances.

A physical analogy of the process of NLM can be given by consideration of a three-dimensional object composed of a set of balls joined together by springs. If the object is pushed onto a flat surface and the tension in the springs allowed to equalize, the result is a two-dimensional representation of a three-dimensional object. The equalization of tension in the springs is equivalent to minimization of the error function in eqn

(4.3). A two-dimensional plot produced by the NLM process has some interesting features. Each axis of the plot consists of some (unknown) non-linear combination of the properties which were used to define the original *N*-dimensional data space. Thus, it is not possible to plot another point directly onto an NLM; the whole map must be recalculated with the new point included. An example of an NLM which includes both training set and test set compounds is shown in Fig. 4.15. This plot was derived from a set of bicyclic amine derivatives which were described by nine calculated parameters. Antiviral activity results were obtained from a plaque-reduction assay against influenza A virus. It can be seen from the map that the active compounds are grouped together in one region of space. Some of the test set compounds lie closer to this region of the plot, though none of them within it, and thus the expectation is that these compounds are more likely to be active than the other members of the test set. This is a good example of the use of an NLM as a means for deciding the order in which compounds should be made or tested. Another use for NLM is to show how well physicochemical property space is spanned by the compounds in the training set, or test set for that matter. Regions of space on the NLM which do not contain points probably indicate regions of *N*-space which do not contain samples. The qualifier 'probably' was used in the last statement because the space on an NLM does not correspond directly to space in *N* dimensions. A map is produced to meet the criterion of the preservation of interpoint distances so as we move about in the 2-space of an NLM this might be equivalent to quite strange moves in *N*-space. Small distances on the NLM may be equivalent to large distances in the space of some variables, small or zero distances with respect to other

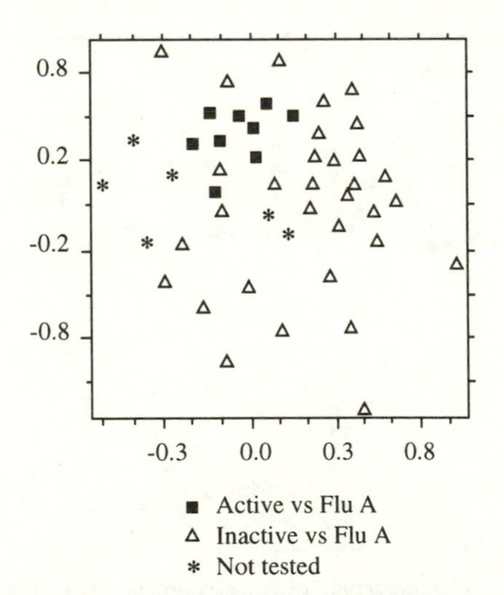

Fig. 4.15. NLM of a set of antiviral bicyclic amine derivatives (from Hudson *et al.* 1989, copyright (1989) Wellcome Foundation).

variables and may even involve a change of direction in the space of some variables.

Another example of the use of NLM to treat chemical data is shown in Fig. 4.16. This NLM was calculated from the same GC–MS data used to produce the principal component scores plot shown in Fig. 4.14. The NLM clearly groups the samples into nine different categories, the

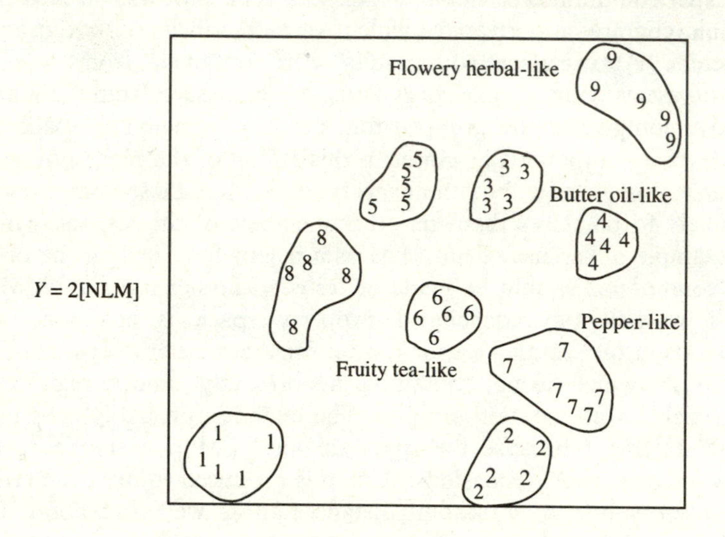

Fig. 4.16. NLM of natural orange aroma samples described by 32 GC–MS peaks (from Lin *et al.* 1993 with permission of Elsevier Science).

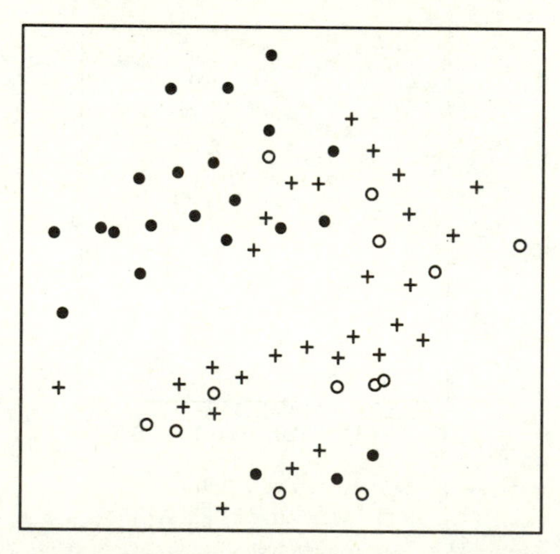

Fig. 4.17. NLM of hallucinogenic phenylalkylamine derivatives described by 24 physicochemical properties; • is active, + is low activity, 0 is inactive (from Clare 1990, copyright (1990) American Chemical Society).

descriptions of the samples are comments made by a human testing panel (see later). Figure 4.17 shows another example of an NLM, this time from the field of drug design. This plot shows 63 hallucinogenic phenylalkylamine derivatives characterized by 24 physicochemical properties. Compounds with high activity are mostly found in the top-left quadrant of the map, the inactive and low-activity compounds being mixed in the rest of the space of the map. Interestingly, this map also shows three active compounds which are separated from the main cluster of actives. These compounds lie quite close to the edge of the plot and thus in a region of the NLM space that might be expected to behave in a peculiar fashion. They may actually be quite similar to the rest of the active cluster, in other words the map may 'join up' at the axes and they are simply placed there as a good compromise in the minimization of the error function. An alternative explanation is that these compounds exert their activity due to some unique features, they may act by a different mechanism or perhaps occupy a different part of the binding site of a biological receptor. Display methods are quite good tools for the identification of compounds, samples, or objects which have different features to the rest of the set.

Figure 4.18 illustrates the use of the power term, ρ, in eqn (4.3). The bicyclic amine data set shown in Fig. 4.15 was mapped using a value of two for this term. With $\rho = 2$, both large and small interpoint distances are equally preserved; this compromise ensures the best overall mapping of the N-space interpoint distances. Figure 4.18 shows the result of mapping this same data set using a value of -2 for ρ. This has the effect of preserving the larger interpoint distances at the expense of the smaller ones; the result is to 'collapse' local clusters of points thus emphasizing the

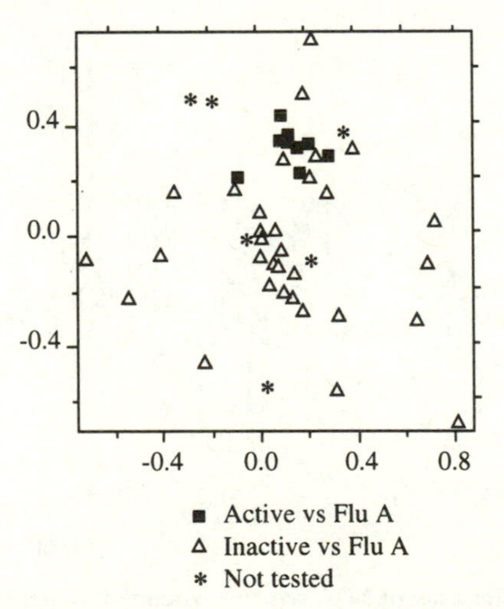

Fig. 4.18. NLM, using a power term $\rho = -2$, of the antiviral bicyclic amine derivatives shown in Fig. 4.15 (from Hudson *et al.* 1989, copyright (1989) Wellcome Foundation).

similarities between compounds. The effect on the data set has been quite dramatic: the active compounds still cluster together and it can be seen that none of the test set compounds join this cluster. However, one of the test set compounds now lies very close to the cluster of actives and thus becomes a much more interesting synthetic target. Two of the remaining test set compounds are close together (only one need be made) and one of the test set compounds has been separated from the rest of the set. This latter compound may now represent an interesting target to make, as it may be chemically different to the rest of the test set, or may be ignored since it lies a long way from the active cluster. Synthetic feasibility and the judgement of the research team will decide its fate.

The final examples of the use of display methods to be shown here also involve a different type of descriptor data, results from a panel of human testers. In the analysis of natural aroma (NOA) samples reported earlier (Lin *et al.* 1993) a human testing panel was trained over a period of three months using pure samples of 15 identified components of the NOA samples. A quantitative descriptive analysis (QDA) report form was devised during the course of the training; the QDA form was used to assign a score to a number of different properties of the NOA samples. PCA of the QDA data for the same samples as shown in Fig. 4.14 resulted in the explanation of 58 per cent of the variance in the data set in the first three PCs. A scores plot on these three PC axes is shown in Fig. 4.19 where it can be seen that the NOA samples are broadly grouped together into different categories, but the classifications are not as tight as those shown

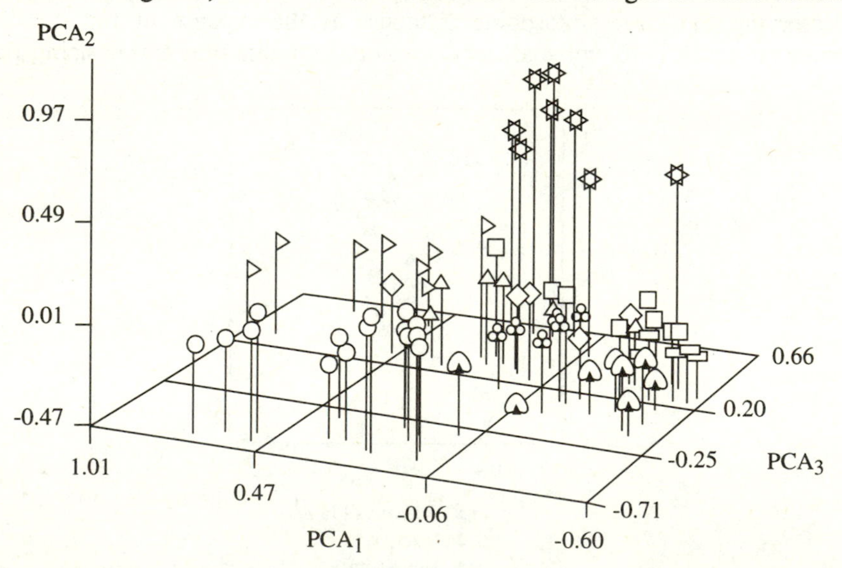

Fig. 4.19. Scores plot for a set of NOA samples described by sensory QDA data. The QDA data was autoscaled and variance-weighted (see reference for details). Symbols are the same as those used in Fig. 4.14 (from Lin *et al.* 1993, with permission of Elsevier Science).

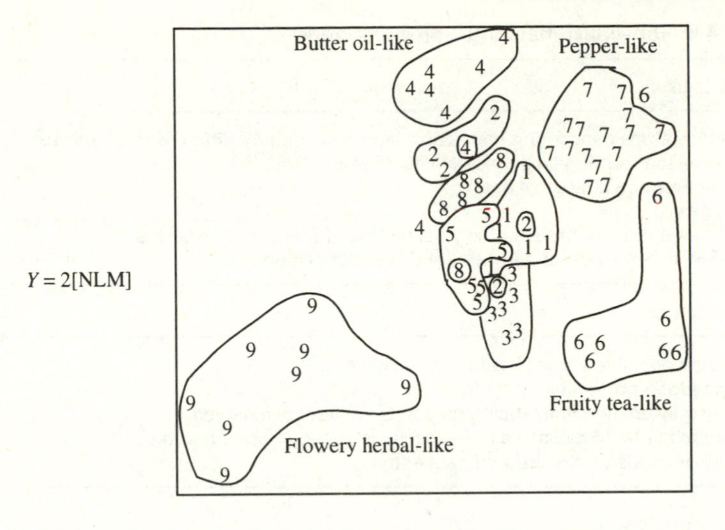

Fig. 4.20 NLM of a set of NOA samples described by Fisher-weighted sensory QDA data. Symbols are the same as those used in Fig. 4.14 (from Lin *et al.* 1993, with permission of Elsevier Science).

in Fig. 4.14. Figure 4.20 shows a non-linear map of Fisher-weighted QDA where it can be seen that some of the categories are quite well separated but not as clearly as the NLM from GC–MS data (see Fig. 4.16).*

Some of the advantages and disadvantages of non-linear mapping as a multivariate display technique are listed in Table 4.5. Most of these have been discussed already in this section but a couple of points have not. Since the technique is an unsupervised learning method, it is unlikely that any grouping of objects will happen by chance. Any cluster of points seen on an NLM generally represents a cluster of points in the *N*-dimensional space. Such groupings may happen by chance although this is much more likely to occur when a supervised learning method, which seeks to find or create patterns in a data set, is employed. The significance of a group of points found on a non-linear map, or any other display for that matter, may be assessed by a method called **cluster significance analysis** as discussed in Chapter 5. The fact that the display is dependent on the order of the compounds and changes as compounds are added or removed is a consequence of the minimization of the error function. The calculated map depends on the initial guess for the 2-space points since the minimizer will find the nearest local minimum rather than the global minimum (if one exists). A common way to choose the initial positions of the points in

* Fisher-weighting and variance-weighting are different procedures for weighting variables according to their ability to classify samples (see Varmuza 1980).

Table 4.5 Non-linear mapping—pros and cons

Advantage
No assumptions concerning mechanism and may identify different mechanisms
Unsupervised learning so chance effects unlikely
Does not require biological data
Non-linear
Can change the emphasis on the preservation of interpoint distances
Can view multivariate data in two (or three) dimensions
Disadvantage
Unknown non-linear combination of variables
Cannot plot a point directly on the map
Display may change dramatically as points are added/removed
Cannot relate NLM distances to *N*-space distances (mapping errors)
Display depends on the order of data entry

2-space is to assign them randomly, but a disadvantage of this is that running the NLM routine several times on the same data set may produce several different maps. One approach to overcoming this problem is to use principal component scores as the initial guess for the 2-space positions; a disadvantage of this is that the resultant map may be more 'linear' than is desirable. Since the error function is calculated over a summation of the distance differences, adding or removing points may alter the subsequent display. This can be disconcerting to newcomers to the method, particularly when we are accustomed to display methods which give only one 'answer'.

4.4 Summary

Multivariate display methods are very useful techniques for the inspection of high-dimensional data sets. They allow us to examine the relationships between points (compounds, samples, etc.) in both training and test sets, and between descriptor variables. Linear and non-linear methods are available, both with advantages and disadvantages, which have proved useful in numerous chemical applications. The linear approach (PCA) forms the basis of a variety of multivariate techniques as described later in this book. Finally, it is not possible to say in advance which, if any, is the best approach to use.

References

Clare, B. W. (1990). *Journal of Medicinal Chemistry*, **33**, 687–702.

Digby, P. G. N. and Kempton, R. A. (1987). *Multivariate analysis of ecological communities*, pp. 19–22. Chapman and Hall, London.

Dizy, M., Martin-Alvarez, P. J., Cabezudo, M. D., and Polo, M. C. (1992). *Journal of the Science of Food and Agriculture*, **60**, 47–53.

Ghauri, F. Y., Blackledge, C. A., Glen, R. C., Sweatman, B. C., Lindon, J. C., Beddell, C. R., *et al.* (1992). *Biochemical Pharmacology*, **44**, 1935–46.

Hudson, B. D., Livingstone, D. J., and Rahr, E. (1989). *Journal of Computer-aided Molecular Design*, **3**, 55–65.

Jackson, J. E. (1991). *A user's guide to principal components*, pp. 155–72. Wiley, New York.

Kowalski, B. R. and Bender, C. F. (1973). *Journal of the American Chemical Society*, **95**, 686–93.

Lewi, P. J. (1986). *European Journal of Medicinal Chemistry*, **21**, 155–62.

Lin, J. C. C., Nagy, S., and Klim, M. (1993). *Food Chemistry*, **47**, 235–45.

Livingstone, D. J. (1991). Pattern recognition methods in rational drug design. In *Molecular design and modelling: concepts and applications, Part B*, Methods in enzymology, Vol. 203, (ed. J. J. Langone), pp. 613–38. Academic Press, San Diego.

Seal, H. (1968). *Multivariate Analysis for Biologists*. Methuen, London.

Schultz, T. W. and Moulton, M. P. (1985). *Bulletin of Environmental Contamination and Toxicology*, **34**, 1–9.

Van de Waterbeemd, H., El Tayar, N., Carrupt, P.-A., and Testa, B. (1989). *Journal of Computer-aided Molecular Design*, **3**, 111–32.

Varmuza, K. (1980). *Pattern recognition in chemistry*, pp. 106–9. Springer-Verlag, New York.

5
Unsupervised learning

5.1 Introduction

The division of topics into chapters is to some extent an arbitrary device to produce manageable portions of text and, in the case of this book, to group together more or less associated techniques. The common theme underlying the methods described in this chapter is that the property that we wish to predict or explain, a biological activity, chemical property, or performance characteristic of a sample, is not used in the analytical method. Oddly enough, one of the techniques described here (nearest-neighbours) does require knowledge of a dependent variable in order to operate, but that variable is not directly involved in the analysis. The display methods described in Chapter 4 are also unsupervised learning techniques, and could have been included in this section, but I felt that display is such a fundamental procedure that it deserved a chapter of its own. Cluster analysis, described in Section 5.4, may also be thought of as a display method since it produces a visual representation of the relationships between samples or parameters. Thus, the division between display methods and unsupervised learning techniques is mostly artificial.

5.2 Nearest-neighbour methods

A number of different methods may be described as looking for nearest neighbours, e.g., cluster analysis (see Section 5.4), but in this book the term is applied to just one approach, *k*-nearest-neighbour. The starting point for the *k*-nearest-neighbour technique (KNN) is the calculation of a distance matrix as required for non-linear mapping. Various distance measures may be used to express the similarity between compounds but the Euclidean distance, as defined in eqn (4.2) (reproduced below), is probably most common:

$$d_{ij} = \sqrt{\left(\sum_{k=1,N} (d_{i,k} - d_{j,k})^2\right)} \tag{5.1}$$

where d_{ij} is the distance between points i and j in N-dimensional space. A distance matrix is a square matrix with as many rows and columns as the number of rows in the starting data matrix. Table 5.1 shows a sample distance matrix from a data set containing ten samples. The diagonal of this matrix consists of zeroes since this represents the distance of each point from itself. The bottom half of the matrix gives the distance, at the intersection of a row and column, between the samples represented by that row and column; the matrix is symmetrical, i.e., distance B→A = distance A→B, so the top half of the matrix is not shown here. An everyday example of a distance matrix is the mileage chart, which can be found in most road atlases, for distances between cities.

Table 5.1 Distance matrix for ten samples

A	0									
B	1.0	0								
C	2.6	2.5	0							
D	2.8	2.6	1.3	0						
E	3.2	2.2	2.8	2.1	0					
F	3.4	2.4	3.1	3.0	1.3	0				
G	3.7	3.4	4.1	3.0	1.3	1.3	0			
H	6.2	5.3	4.3	3.0	3.0	3.2	2.9	0		
I	9.8	9.7	4.0	3.7	6.2	7.5	6.2	3.5	0	
J	10.0	9.9	4.4	4.0	6.3	7.6	6.4	3.6	1.2	0
	A	B	C	D	E	F	G	H	I	J

The classification of any unknown sample in the distance matrix may be made by consideration of the classification of its nearest neighbour. This involves scanning the row and column representing that sample to identify the smallest distance to other samples. Having identified the distance (or distances) it is assumed that the classification of the unknown will be the same as that of the nearest neighbour, in other words samples that are similar in terms of the property space from which the distance matrix was derived will behave in a similar fashion. This mimics the 'common-sense' reasoning that is customarily applied to the interpretation of simple two-dimensional plots, the difference being that here the process is applied in N-dimensions. Figure 5.1 shows a two-dimensional representation of this process. The training set compounds are shown marked as A and I for active and inactive; the unknown, test set, compounds are indicated as X, Y, and Z. The nearest neighbour to compound X is active and that to compound Y is inactive, and this is how these two would be classified on the basis of one nearest neighbour. Classification for compound Z is more difficult as its two apparently equidistant neighbours have different activities. Although one of these neighbours may be slightly closer when the values in the distance matrix are examined, it is clear that this represents an ambiguous prediction. With the exception of one close active compound the remaining neighbours of compound Z are inactive and the

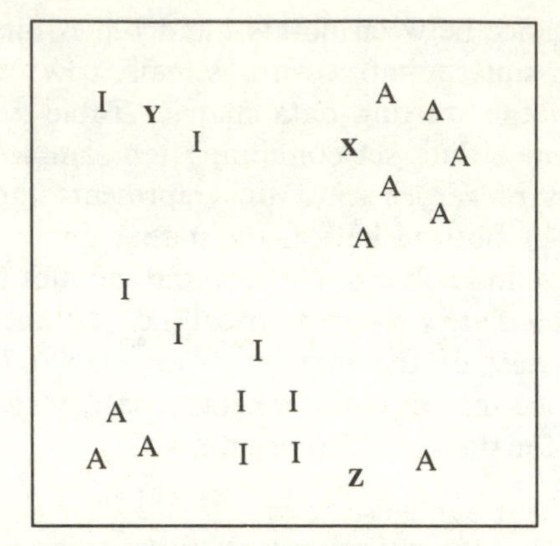

Fig. 5.1. Two-dimensional representation of the KNN technique; training set compounds are represented as A for active and I for inactive, test set compounds as X, Y, and Z.

common-sense prediction would be for Z to be inactive. This is the meaning of the term k in k-nearest-neighbour, k refers to the number of neighbours that will be used for prediction. The choice of a value for k will be determined by the training set; this is achieved by comparing the predictive performance of different values of k for the training set compounds. Figure 5.1 also illustrates a quite common situation in the analysis of multivariate data sets; the two activity classes are not linearly separable (i.e., it is not possible to draw a single straight line that will divide up the space into two regions containing only active or inactive compounds). Some analytical methods operate by the construction of a hyperplane, the multivariate analogue of a straight line, between classes of compounds (see Chapter 7). In the case of a data set such as this, the KNN method will have superior predictive ability.

Nearest-neighbour methods are also able to make multi-category predictions of activity; training set samples can be ranked into any number of classifications, but it is important to maintain the balance between the number of classes and the number of members within a class. Ideally, each class should contain about the same number of members although in some situations (such as where the property is definitely YES/NO) this may not be possible to achieve. The reason for maintaining similar numbers of members in each class is so that a 'random' prediction of membership for any class is not significantly greater than that for the other classes. This raises the question of how to judge the prediction success of a nearest-neighbour method. When a training set is split into two classes, there is a 50 per cent chance of making a successful prediction for any compound, given equal class membership. A success rate of 80 per cent for the training set may sound impressive but is, in fact, little over half as good again as would be expected from purely random guesses. Where classes do differ significantly in size it is possible to change the random expectation in order to judge success. For example, if class 1 contains twice as many members as class 2, the random expectations are ~66 and ~33 per cent respectively. Another aspect of the balance between the number of classes and the size of class membership concerns the dimensionality of the data set. When the number of dimensions in a data set is close to or greater than the number of samples in the set, it is possible to discover linear separation between classes by chance. The risk of this happening is obviously greatest for supervised learning methods (which are intended to find linear separations) but may also happen with unsupervised techniques. There are no 'rules' concerning the relationship between dimensions and data points for unsupervised learning methods but, as with most forms of analysis, it is desirable to keep the problem as simple as possible.

Now for some examples of the application of nearest-neighbour methods to chemical problems. An early example involved the classification of compounds described by calculated NMR spectra (Kowalski and Bender 1972). The data set consisted of 198 compounds divided into three classes (66 each) of molecules containing $CH_3CH_2CH_2$, CH_3CH_2CH, or CH_3CHCH. The NMR spectra were preprocessed (see reference for details) to give 12 features describing each compound and the data set was split in half to give training and test sets. Table 5.2 shows the results for training set and test set predictions using 1-nearest neighbour. Since there are three classes with equal class membership, the random expectation would be 33 per cent correct, and thus it appears that the nearest-neighbour technique has performed very well for this set. Also shown in Table 5.2 are the results for a method called the linear learning machine (see Section 7.2.1) which has performed quite poorly with a success rate only

Table 5.2 Nearest-neighbour classification for NMR (12 features) data set (from Kowalski and Bender 1972, copyright (1972) American Chemical Society)

	Correct / Total	
	Training set	Test set
1-nearest-neighbour		
Class 1	60/66	60/66
Class 2	60/66	60/66
Class 3	64/66	64/66
Total	184/198 = 93%	184/198 = 93%
Learning machine		
Class 1	56/66	54/66
Class 2	12/66	12/66
Class 3	24/66	23/66
Total	92/198 = 46%	89/198 = 45%

slightly above that expected by chance, suggesting that the data is not linearly separable. An application of KNN in medicinal chemistry was reported by Chu and co-workers (1975). In this case the objective was to predict the antineoplastic activity of a test set of 24 compounds in a mouse brain tumour system. The training set consisted of 138 structurally diverse compounds which had been tested in the tumour screen. The compounds, test set, and training set, were described by a variety of sub-structural descriptors giving a total of 421 parameters in all. Various procedures were adopted (see reference for details) to reduce this to smaller sized sets and KNN was employed to make predictions using the different data sets. The KNN predictions were averaged over these data sets to give an overall success rate of 83 per cent. A comparison of the predictions with the experimental results is shown in Table 5.3

Table 5.3 Comparison of predicted and observed antineoplastic activities (from Chu *et al.* 1975, copyright (1975) American Chemical Society)

	Non-active	False negative	Active	False positive
KNN	14	1	6	3
Experimental	17		7	

Scarminio and colleagues (1982) reported a comparison of the use of several pattern recognition methods in the analysis of mineral water samples characterized by the concentration of 18 elements, determined by atomic absorption and emission spectrometry. The result of the application of cluster analysis and SIMCA to this data set is discussed elsewhere (Sections 5.4 and 7.2.2); KNN results are shown in Table 5.4. The performance of KNN in this example is really quite impressive; for two regions, the training set samples are completely correctly classified up to five nearest neighbours and the overall success rate is 95 per cent or better

Table 5.4 KNN classification results for water samples, collected from four regions, described by the concentration of four elements (Ca, K, Na, and Si) (from Scarmino *et al.* 1982, with kind permission)

Region	Number of samples	Number of points incorrectly classified			
		1-NN	3-NN	5-NN	7-NN
Serra Negra	46	2	3	3	2
Lindoya	24	0	0	0	1
São Jorge	7	1	2	1	1
Valinhos	39	0	0	0	0
	Correct (%)	97.3	95.5	96.4	96.4

($\sim$25 per cent success rate for the random expectation). A test set of seven samples was analysed in the same way and KNN was found to classify all the samples correctly, considering up to nine nearest neighbours.

A recent example of the use of KNN in chemistry was reported by Goux and Weber (1993). This study involved a set of 99 saccharide residues, occurring as a monosaccharide or as a component of a larger structure, described by 19 experimentally determined NMR parameters. The NMR measurements included a coupling constant and both proton and carbon chemical shifts. The aim of the work was to see if the NMR data could be used to classify residues in terms of residue type and site of glycoside substitution to neighbouring residues. A further aim was the identification of important NMR parameters, in terms of their ability to characterize the residues. To this end, a number of subsets of the NMR parameters were created and the performance of KNN predictions, in this case 1-NN, were assessed. Table 5.5 illustrates the results of this 1-NN classification for the full dataset and five subsets. For the full dataset of 19 variables, one residue is misassigned, and for two subsets this misclassification is eliminated. Two of the other subsets give a poorer classification rate, demonstrating that not only can the NMR data set be used to classify the residues but also that the important parameters can be recognized.

Table 5.5 Nearest-neighbour classification of glycosides (from Goux and Weber 1993, with permission of Elsevier Science)

Data set	Number misassigned (% correct)
UAR-19 (full set)	1 (99)
DP-12	0 (100)
DP-9	8 (90)
SP-6/1	0 (100)
SP-6/4	1 (99)
SP-6/5	3 (96)

5.3 Factor analysis

Principal component analysis (PCA), as described in Chapter 4, is an unsupervised learning method which aims to identify principal components, combinations of variables, which 'best' characterize a data set. Best here means in terms of the information content of the components (variance) and that they are orthogonal to one another. Each principal component (PC) is a linear combination of the original variables as shown in eqn (4.1) and repeated below

$$PC_q = a_{q,1}v_1 + a_{q,2}v_2 + \ldots a_{q,N}v_N \tag{5.2}$$

The principal components do not (necessarily) have any physical meaning, they are simply mathematical constructs calculated so as to comply with the conditions of PCA of the explanation of variance and orthogonality. Thus, PCA is not based on any statistical model. Factor analysis (FA), on the other hand, is based on a statistical model which holds that any given data set is based on a number of factors. The factors themselves are of two types; common factors and unique factors. Each variable in a data set is composed of a mixture of the common factors and a single unique factor associated with that variable. Thus, for any variable X_i, in an N-dimensional data set we can write

$$X_i = a_{i,1}F_1 + a_{i,2}F_2 + \ldots a_{i,p}F_p + E_i \tag{5.3}$$

where each $a_{i,j}$ is the loading of variable X_i on factor F_j, and E_i is the residual variance specific to variable X_i. The residual variance is also called a unique factor associated with that variable, the common factors being F_1 to F_p which are associated with all variables, hence the term common factors. The similarity with PCA can be seen by comparison of eqns (5.3) and (5.2). Indeed, PCA and FA are often confused with one another and since the starting point for a FA can be a PCA this is perhaps not surprising.

The two methods, although related, are different. In the description of eqn (5.3) the loadings, $a_{i,j}$, were described as the loadings of variables X_i on factor F_j so as to point out the similarity to PCA. Expression of this equation in the (hopefully) more familiar terms of PCA gives these loadings as the loadings of each of the common factors, F_1 to F_p, onto variable X_i. In other words, PCA identifies principal components which are linear combinations of the starting variables; FA expresses each of the starting variables as a linear combination of common factors. PCA seeks to explain all of the variance in a data set; FA seeks to factor (hence the name) the variance in a data set into common and unique factors. The unique factors are normally discarded, since it is usually assumed that they represent some 'noise' such as experimental error, and thus FA will reduce the variance of a data set by (it is hoped) removing irrelevant information. Since the unique factors are removed, the remaining common factors all contain variance from at least two, if not more, variables. Common factors are explaining covariance and thus FA is a method which describes covariance, whereas PCA preserves and describes variance. Like PCs, the factors are orthogonal to one another and various rotations (like varimax, see Section 4.2) can be applied in order to simplify them. One of the advantages claimed for FA is that it is based on a 'proper' statistical model, unlike PCA, and that by discarding unique factors the data set is 'cleaned up' in terms of information content. FA, however, relies on a

Table 5.6 Sorted rotated factor loadings (pattern) from factor analysis of meat and fish data* (from Li-Chan *et al.* 1987 with permission of the Institute of Food Technologists)

	Factor 1	Factor 2	Factor 3
Dispersibility	−0.959	0.000	0.000
Solubility	−0.939	0.000	0.000
ANS	0.864	0.000	0.255
Gel-M	0.862	0.000	0.000
Gel-E	0.853	0.288	0.000
EC	−0.848	0.000	0.000
CPA	0.844	0.000	0.000
FBC	−0.604	0.293	0.000
Cookloss	0.529	0.000	−0.457
Protein	0.000	0.916	0.000
Mince-pH	0.000	−0.852	0.285
S.E.P.	0.000	0.821	0.000
SH	0.000	0.617	0.000
Moisture	0.000	0.000	0.966
Fat	0.000	−0.382	−0.891

* Loadings less than 0.2500 have been replaced by zero.

number of assumptions and these are equally claimed as disadvantages to the technique. Readers interested in further discussion of FA and PCA should consult Chatfield and Collins (1980), Malinowski (1991), or Jackson (1991).

What about applications of FA? An interesting example was reported by Li-Chan and co-workers (1987) who investigated the quality of hand-deboned and mechanically deboned samples of meat and fish. The samples were characterized by physicochemical properties such as pH, fat, and moisture content, and by functional properties such as gel strength and per cent cookloss (in terms of weight). Factor analysis of the overall data set of 15 variables for 230 samples extracted three factors which described 70 per cent of the data variance. The factor loadings are shown in Table 5.6 where it can be seen that factor 1 includes the hydrophobic/hydrophilic properties of the salt-extractable proteins, factor 2 describes total and salt-extractable proteins and mince pH, and factor 3 is associated mainly with moisture and fat. The factor loadings may be plotted against one another as for PC loadings (Fig. 4.12) in order to show the relationships between variables. Rotated factor loadings from Table 5.6 are shown in Fig. 5.2 in which various groupings of associated variables may be seen. For example, total and salt extractable protein are associated, as are solubility, dispersibility, and emulsifying capacity. Factor scores may be calculated for the samples and plotted on the factor axes. Figure 5.3 shows factor scores for factor 3 versus factor 2 where it can be seen that the fish samples are quite clearly distinguished from the meat samples which, in turn, fall into two groups, hand-deboned and mechanically deboned.

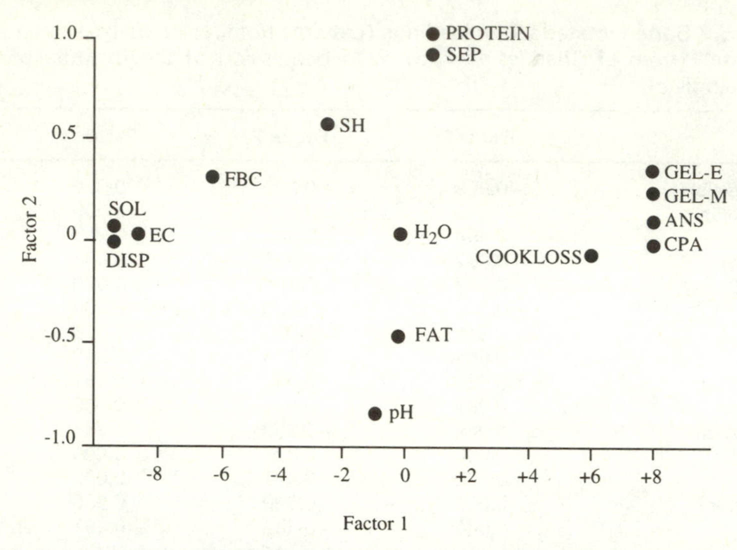

Fig. 5.2. Loadings plot on the first two factors for 15 variables used to describe samples of meat and fish. SOL, solubility; EC, emulsifying capacity; DISP, dispersability; FBC, fat binding capacity; SH, sulphydryl content; PROTEIN, protein content; SEP, salt-extractable protein; H2O, moisture content; pH, pH of a mince suspension; FAT, crude fat content; COOKLOSS, percentage weight lost after cooking; GEL-M, gel strength of mince; GEL-E, gel strength of extract; ANS and CPA, protein surface hydrophobicity using aromatic (ANS) or aliphatic (CPA) fluorescent probe (from Li-Chan *et al.* 1987, with permission of the Institute of Food Technologists).

Takagi and co-workers (1989) applied FA to gas chromatography retention data for 190 solutes measured using 21 different stationary phases. Three factors were found to be sufficient to explain about 98 per cent of the variance of the retention data; physicochemical meanings to these factors were ascribed as shown below:

Factor 1: size
Factor 2: polarity
Factor 3: hydrogen-bonding tendency

As is usually the case with PCA, the attribution of any physical meaning to factors is not straightforward, particularly for the 'later' factors (smaller eigenvalues, less variance explained) from an analysis; this was the case for the third factor. Factor loadings for the three factors are shown in Table 5.7. Part of the argument in favour of factor 3 as a hydrogen-bonding factor is the negative loading of a proton donor stationary phase (HCM 18) and the positive loading of proton acceptor phases (DEGA, PPE5R, and EGA). Another part of the argument is that non-polar stationary phases have approximately zero loadings with this factor.

The attempted physicochemical interpretation of the factors highlights a common problem with PCA and FA, along with the question of how

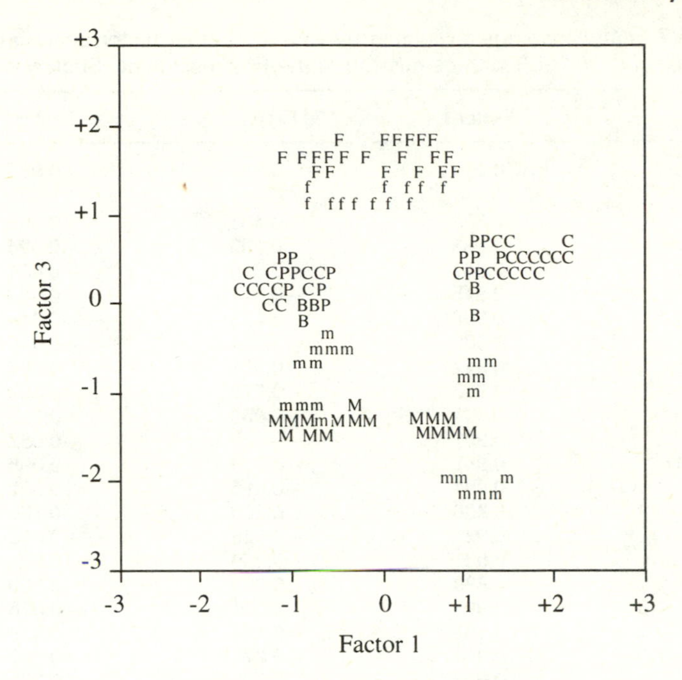

Fig. 5.3. Plot of factor scores for meat and fish samples. C, B, P- hand-deboned chicken, beef, and pork; M, m, mechnically debonded pork and chicken; F, cod fish; f, cod fish in presence of cryoprotectants (from Li-Chan *et al.* 1987, with permission of the Institute of Food Technologists).

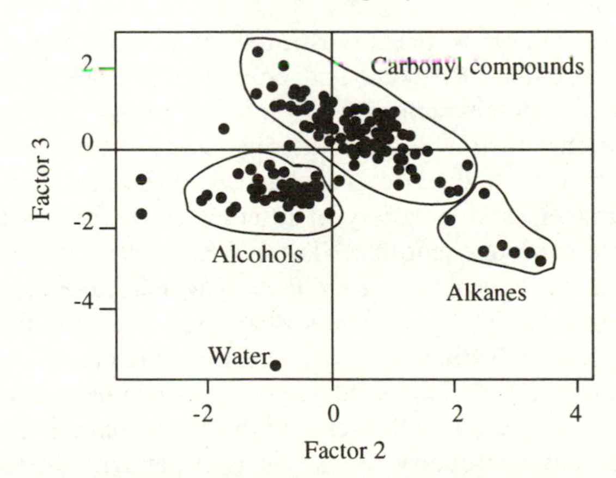

Fig. 5.4. Scores plot on factors 3 and 2 derived from GC retention data (from Takagi *et al.* 1989, with permission of the Pharmaceutical Society of Japan).

significant is a factor (or PC) which only describes a few per cent of the data variance. The significance of factor 3 is uncertain but it is clearly useful since a scores plot of factor 3 versus factor 2 separates the solutes in terms of chemical functionality as shown in Fig. 5.4. The physicochemical

Table 5.7 Factor loadings calculated by PFA (principal factor analysis) method (from Takagi *et al.* 1989 with permission of the Pharmaceutical Society of Japan)

	Factor 1	Factor 2	Factor 3
CCR*	0.922	0.990	0.997
AP1-L	0.898	0.431	−0.001
CASTOR	0.983	0.133	−0.094
CW1000	0.947	−0.315	−0.004
DEGA	0.945	−0.311	0.083
D2EHS	0.968	0.238	−0.044
DIDP	0.969	0.224	−0.040
DC550	0.942	0.321	0.087
EGA	0.953	−0.279	0.089
HCM18	0.971	0.050	−0.227
HYP	0.901	−0.419	−0.052
IGE880	0.981	−0.183	−0.005
NPGA	0.994	−0.095	0.022
PPE5R	0.966	0.209	0.150
QUAD	0.931	−0.326	−0.089
SE30	0.931	0.359	0.000
SAIB	0.998	0.002	0.030
TCP	0.993	0.031	−0.016
TX305	0.982	−0.181	−0.055
U2000	0.994	−0.071	−0.037
VF50	0.912	0.399	0.020
XF1150	0.960	−0.195	0.138

* Cumulative contribution ratio.

interpretation of factors is nicely illustrated by a factor analysis of solvent parameters reported by Svoboda and co-workers (1983). Many attempts have been made to characterize solvents in terms of their effect on chemical reactions, their ability to dissolve solutes, their effect on properties such as spectra, and so on. This has led to the development of many different parameters and a variety of attempts have been made to relate these parameters to one another. Table 5.8 shows the loadings of 20 parameters, for 51 solvents, on four factors which have been rotated by the varimax method. The parameters associated with the first factor describe electrophilic solvation ability, while those associated with factor 2 concern solvent polarity. The third factor is associated with nucleophilic solvation ability and the fourth factor with dispersion solvation forces. It was proposed that a property, A, which is dependent on solvent effects could be described by an equation consisting of these four factors as shown in eqn (5.4):

$$A = A_0 + aAP + bBP + eEP + pPP \tag{5.4}$$

where AP (acidity parameter) is the electrophilic factor, BP (basicity parameter) is the nucleophilic parameter, EP (electrostatic parameter) the

Table 5.8 Coordinates of parameters in factor space of the selected set (set 2) (from Svoboda *et al.* 1983 with kind permission)

Parameter	F_1	F_2	F_3	F_4
B	−0.079	0.012	0.610	0.008
$E_T(30)$	0.181	0.263	0.002	0.092
Z	0.305	−0.018	0.021	0.133
S_1[a]	0.261	0.078	0.144	0.014
S_2[b]	0.225	0.208	−0.010	0.051
DN	0.187	−0.208	0.560	−0.161
ε	0.154	0.237	0.026	−0.047
n_D^{20}	−0.010	0.044	−0.010	−0.600
YP[c]	−0.097	0.473	0.090	0.068
pp[d]	−0.014	0.046	−0.008	−0.604
E	0.297	0.073	−0.042	0.124
a^{14_N}	0.355	−0.015	0.022	0.037
AN	0.390	−0.066	−0.026	−0.014
π^*	0.026	0.413	−0.032	−0.250
log P	−0.113	−0.103	−0.152	−0.167
δ	0.373	0.024	−0.014	−0.114
χ_R	0.094	−0.480	−0.046	0.106
δ^2	0.363	−0.011	−0.071	−0.113
β	−0.137	0.169	0.472	0.140
$^n\chi$[e]	0.028	−0.320	0.159	−0.222

[a] S_1, the parameter S defined by Zelinski.
[b] S_2, the parameter S defined by Brownstein.
[c] The Kirkwood function of dielectric constant $YP = (\varepsilon - 1)/(2\varepsilon + 1)$.
[d] The function of refractive index $PP = (n^2 - 1)/(n^2 + 1)$.
[e] The index of molecular connectivity of the nth order.

polar factor, and *PP* (polarizability parameter) the dispersion factor. A_0 is the value of the solvent-dependent property in a medium in which the solvent factors are zero (cyclohexane was suggested as a suitable solvent for this). The coefficients *a*, *b*, *e*, and *p* are fitted by regression to a particular data set and represent the sensitivity of a process, as measured by the values of *A*, to the four solvent factors. Application of this procedure to 22 chemical data sets identified examples of processes with quite different dependencies on these solvent properties.

The final example of FA to be discussed here involves a number of insecticides, which are derivatives of the pyrethroid skeleton shown in Fig. 5.5. Computational chemistry methods were used to calculate a set of 70

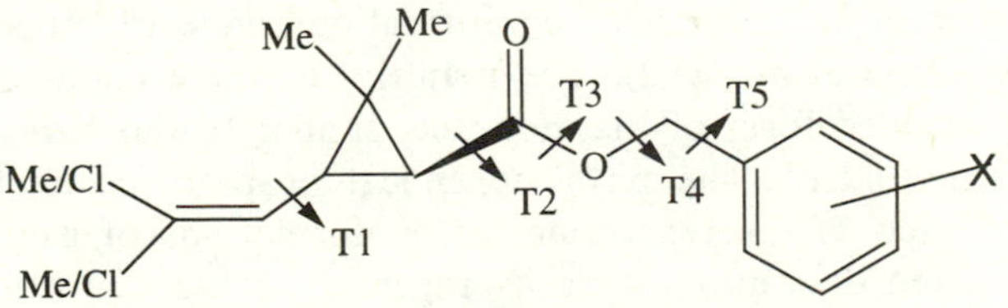

Fig. 5.5. Pyrethroid parent structure; rotatable bonds are indicated by arrows (from Hudson *et al.* 1992, copyright (1992) Wellcome Foundation Ltd).

Table 5.9 The molecular features of the QSAR pyrethroids (identified by FA) (from Ford *et al.* 1989 with permission of the Society of Chemical Industry)

Factor	Principal associated descriptors and loadings	Molecular feature
1	A11(0.97), A12(0.99), A13(0.91), A16(−0.99), A17(−0.95), MW(−0.63)	The nature of the acid moiety indicated by associated MW and partial atomic charges
2	NS9(0.97), NS10(0.97), NS8(0.96), NS11(0.96), ES15(0.77), NS7(0.77)	Tendency of the atoms around the central ester linkage to accept and the *cis* geminal methyl to donate electrons
3	ES1(0.94), ES7(0.94), ES8(0.94), ES9(0.94), ES10(0.95)	Tendency of the atoms associated with the ester linkage to donate electrons
4	A3(0.84), A5(0.84), A10(0.85), ET(−0.84)	Partial atomic charges on the *meta* carbon atoms of the benzyl ring and the carbonyl carbon
5	NS2(0.90), NS3(0.75), NS5(0.77), NS6(0.87), A7(−0.77)	Tendency of the *ortho*- and *meta*-carbon atoms of the benzyl ring to accept electrons
6	DCA(0.86), SA(0.79), CD(0.71), VWV(0.71)	Molecular bulk, surface area and distance of closest approach
7	DVZ(0.82), DM(0.81)	Dipole strength and orientation
8	MW(0.70)	Molecular weight due to the alcohol moiety

molecular properties to describe these compounds (Ford *et al.* 1989). Factor analysis identified a set of eight factors (Table 5.9) which explained 99 per cent of the variance in the chemical descriptor set. The physicochemical significance of the factors can be judged to some extent by an examination of the properties most highly associated with each factor, as shown in the table. The factors were shown to be of importance in the description of several biological properties of these compounds (see Section 8.5). These pyrethroid analogues are flexible as indicated by the rotatable bonds marked in Fig. 5.5. The physicochemical properties used to derive the factors shown in Table 5.9 were based on calculations carried out on a single conformation of each compound using a template from an X-ray crystal structure. In an attempt to take account of conformational flexibility, molecular dynamics simulations were run on the compounds and a number of representative conformations were obtained for each analogue (Hudson *et al.* 1992). The majority of these conformations represent an extended form of the molecule, similar to the X-ray template, but some are 'folded'. The physicochemical property calculations were repeated for each of the representative conformations of each analogue and the resulting descriptors were averaged. Running factor analysis on this time-averaged set resulted in the identification of nine 'significant' factors (eigenvalues greater than 1), one more than the factor analysis of the static set. This additional factor suggests that there is extra information in the time-averaged set. Several of the static and time-averaged factors were highly correlated with one another and it was shown that these factors could be used to explain the lifetimes of the folded conformations.

5.4 Cluster analysis

Chatfield and Collins (1980), in the introduction to their chapter on cluster analysis, quote the first sentence of a review article on cluster analysis by Cormack (1971): 'The availability of computer packages of classification techniques has led to the waste of more valuable scientific time than any other "statistical" innovation (with the possible exception of multiple-regression techniques).' This is perhaps a little hard on cluster analysis and, for that matter, multiple regression but it serves as a note of warning. The aim of this book is to explain the basic principles of the more popular and useful multivariate methods so that readers will be able to understand the results obtained from the techniques and, if interested, apply the methods to their own data. This is *not* a substitute for a formal training in statistics; the best way to avoid wasting one's own valuable scientific time is to seek professional help at an early stage.

Cluster analysis (CA) has already been briefly mentioned in Section 2.3, and a dendrogram was used to show associations between variables in Section 3.4. The basis of CA is the calculation of distances between objects in a multidimensional space using an equation such as eqn (5.1). These distances are then used to produce a diagram, known as a dendrogram, which allows the easy identification of groups (clusters) of similar objects. Figure 5.6 gives an example of the process for a very simple two-dimensional data set. The two most similar (closest) objects in the two-dimensional plot in part (a) of the figure are A and B. These are joined together in the dendrogram shown in part (b) of the figure where they have a low value of dissimilarity (distance between points) as shown on the scale. The similarity scale is calculated from the interpoint distance matrix by finding the minimum and maximum distances, setting these equal to some arbitrary scale numbers (e.g., 0 and 1), and scaling the other distances to lie between these limits. The next smallest interpoint distance is between point C and either A or B, so this point is joined to the A/B cluster. The next smallest distance is between D and E so these two points form a cluster and, finally, the two clusters are joined together in the dendrogram. This process is hierarchical and the links between clusters have been single; the procedure is known, unsurprisingly, as single-link hierarchical cluster analysis and is one of the most commonly used methods. Another point to note from this description of CA is that clusters were built up from

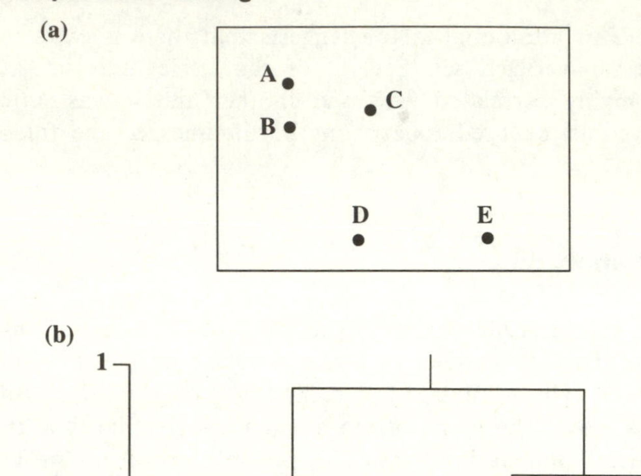

Fig. 5.6. Illustration of the production of a dendrogram for a simple two-dimensional data set.

individual points, the process is agglomerative. CA can start off in the other direction by taking a single cluster of all the points and splitting off individual points or clusters, a divisive process.

There are many different ways in which clusters can be generated; all of the examples that will be described in this section use the agglomerative, hierarchical, single-linkage method, usually referred to as 'cluster analysis'. Most textbooks of multivariate analysis have a chapter describing some of the alternative methods for performing CA, and Willett (1987) deals with chemical applications. It may have been noticed that in this description of CA the points to be clustered were referred to as just that, points in a multidimensional space. They have not been identified as samples or variables since CA, like many multivariate methods, can be used to examine relationships between samples or variables. For the former we can view the data set as a collection of p objects in an N-dimensional parameter space. For the latter we can imagine a data set 'turned on its side' so that it is a collection of N objects in a p-dimensional sample space. When using CA to examine the relationships between variables, the distance measure employed is often the correlation coefficients between variables.

The study of mineral waters characterized by elemental analysis discussed in Section 5.2 (Scarminio *et al.* 1982) provides a nice example of the use of CA to classify samples. Figure 5.7 shows a dendrogram of water samples from one geographical region (Lindoya) described by the concentrations of four elements. The water samples were drawn from six

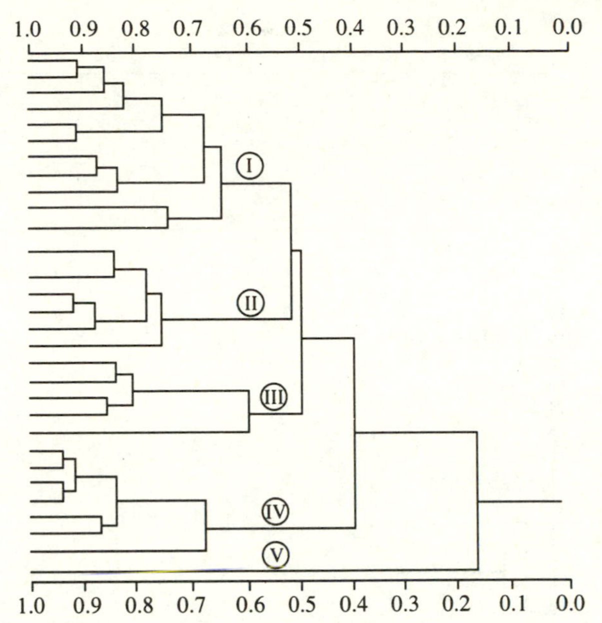

Fig. 5.7. Dendrogram of water samples characterized by their concentrations of Ca, K, Na, and Si (from Scarminio *et al.* 1982, with kind permission).

different locations in this region and one group on the dendrogram, cluster IV, contained all the samples from one of these locations. The samples from the other five locations are contained in clusters I, II, and III. One sample, cluster V, is clearly an outlier from this set and thus must be subject to suspicion.

The characterization of fruit juices by various analytical measurements was used as an example of a principal component scores plot (Fig. 4.9) in Chapter 4 (Dizy *et al.* 1992). A dendrogram from this data is shown in Fig. 5.8 where it is clearly seen that the grape, apple, and pineapple juice samples form distinct clusters. The apple and pineapple juice clusters are grouped together as a single cluster which is quite distinct from the cluster of grape juice samples. This is interesting in that it mimics the results of the PCA; on the scores plot, all three groups are separated, but the first component mainly serves to separate the grape juices from the others while the second component separates apple and pineapple juices. This is a good illustration of the way that different multivariate methods tend to produce complementary and consistent views of the same data set.

The dendrogram in Fig. 5.9 is derived from a data matrix of ED_{50} values for 40 neuroleptic compounds tested in 12 different assays in rats (Lewi 1976). This is an example of a situation in which the data involves multiple dependent variables (see Chapter 8), but here the multiple biological data is used to characterize the tested compounds. The figure

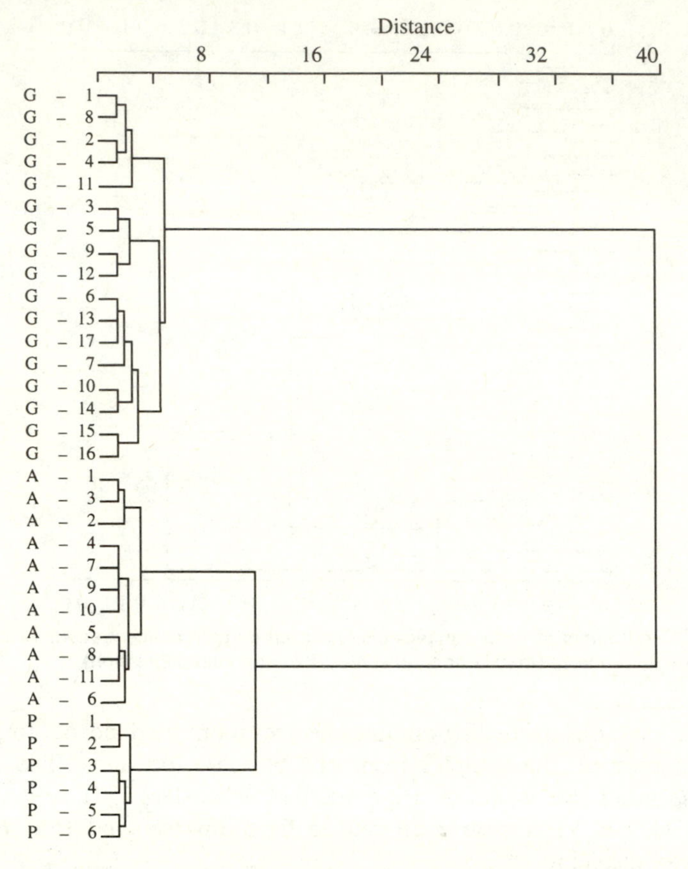

Fig. 5.8. Dendrogram showing the associations between grape (G), apple (A), and pineapple (P) juice samples described by 15 variables (from Dizy *et al.* 1992, with permission of the Society of Chemical Industry).

demonstrates that the compounds can be split up into five clusters with three compounds falling outside the clusters. Compounds within a cluster would be expected to show a similar pharmacological profile and, of course, there is the finer detail of clusters within the larger clusters. A procedure such as this can be very useful when examining new potential drugs. If the pharmacological profile of a new compound can be matched to that of a marketed compound, then the early clinical investigators may be forewarned as to the properties they might expect to see.

The final example of a dendrogram to be shown here, Fig. 5.10, is also one of the largest. This figure shows one thousand conformations of an insecticidal pyrethroid analogue (see Fig. 5.5) described by the values of four torsion angles (Hudson *et al.* 1992). A dendrogram such as this was used for the selection of representative conformations from the one thousand conformations produced by molecular dynamics simulation.

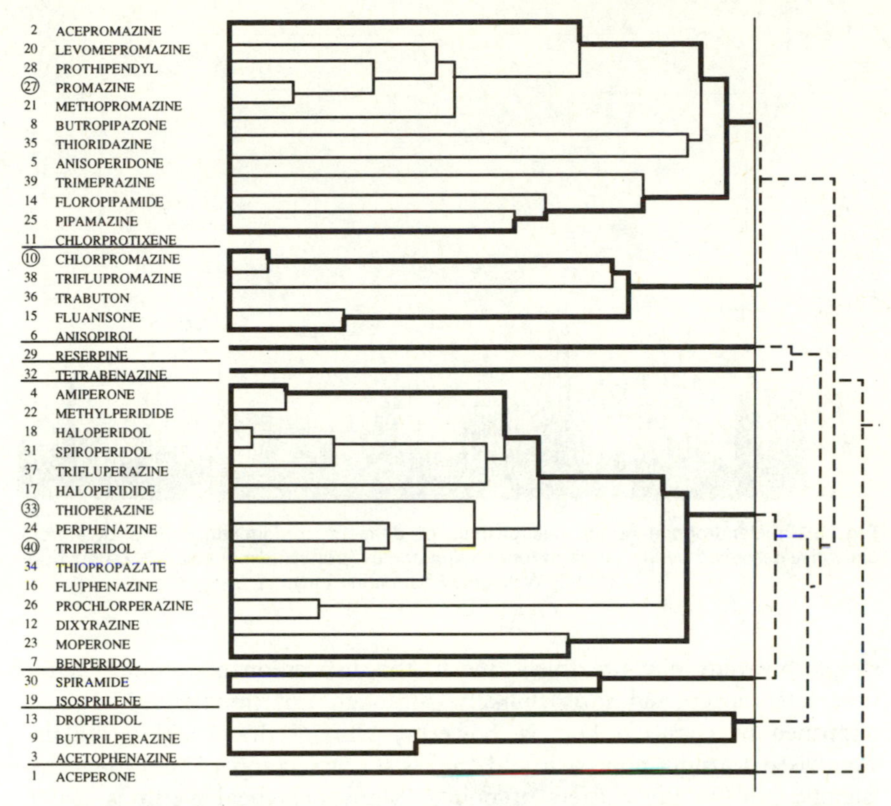

Fig. 5.9. Dendrogram of the relationships between neuroleptic drugs characterized by 12 different biological tests (from Lewi 1976, with permission of Editio Cantor Verlag).

Conformations were chosen at equally spaced intervals across the dendrogram ensuring an even sampling of the conformational space described by the torsion angles. In fact, the procedure is not as simple as this and various approaches were employed (see reference for details) but sampling at even intervals was shown to be suitable.

5.5 Cluster significance analysis

The advantage of unsupervised learning methods is that any patterns that emerge from the data are dependent on the data employed. There is no intervention by the analyst, other than to choose the data in the first place, and there is no attempt by the algorithm employed to 'fit' a pattern to the data, or seek a correlation, or produce a discriminating function (see Chapter 7). Any groupings of points which are seen on a non-linear map, a principal components plot, a dendrogram, or even a

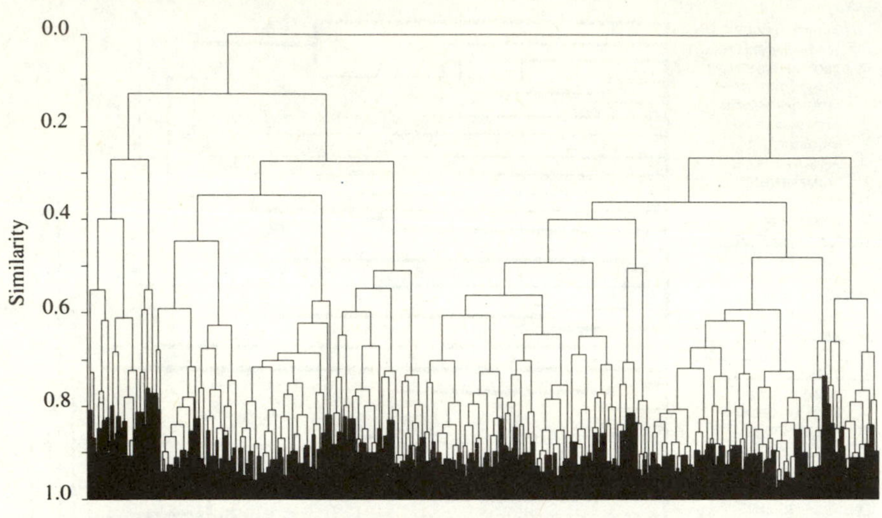

Fig. 5.10. Dendrogram of the relationships of different conformations of a pyrethroid derivative described by the values of four torsion angles (from Hudson *et al.* 1992, copyright (1992) Wellcome Foundation Ltd).

simple bivariate plot are solely due to the disposition of samples in the parameter space and it is unlikely, although not impossible, to have happened by chance. There is, however, a major drawback to the unsupervised learning approach and that is an evaluation of the quality or 'significance' of any clusters of points. Many analytical methods, particularly the parametric techniques based on assumptions about population distributions, have significance tests built in. If we look at the principal component scores plot for the fruit juices (Fig. 4.9) or the dendrogram for the same data (Fig. 5.8) it seems obvious that the groupings have some 'significance', but is this always the case? Is it possible to judge the quality of some unsupervised picture? McFarland and Gans (1986) addressed this problem by means of a method which they termed *cluster significance analysis* (CSA). The concept underlying this method is quite simple: for a given display of N samples which contains a cluster of M active (or otherwise interesting) samples, how 'tight' is the cluster of M samples compared with all the other possible clusters of M samples? Various measures of tightness could be used but the one chosen was the mean squared distance (MSD) which involves taking the sum of the squared distances between each pair of points in the cluster divided by the number of points in the cluster (M).

The process is nicely illustrated by a hypothetical example from the original report. Figure 5.11 shows a two-dimensional plot of six compounds, three active and three inactive. The total squared distance (TSD) for the active cluster is given by

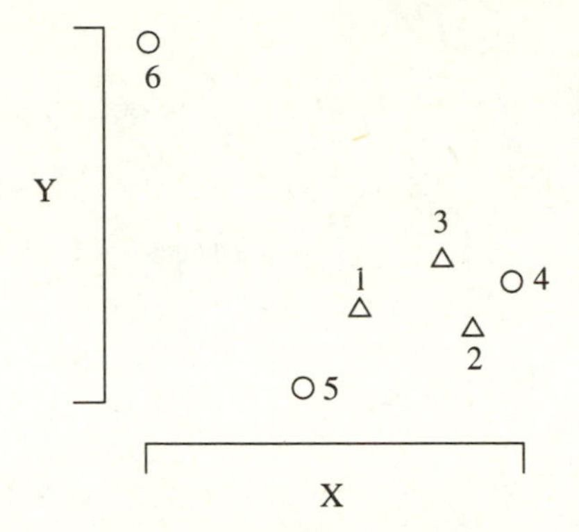

Fig. 5.11. Plot of active (Δ) and inactive (O) compounds described by two parameters (from McFarland and Gans 1986, copyright (1986) American Chemical Society).

$$TSD = (x_1-x_2)^2 + (y_1-y_2)^2 + (x_1-x_3)^2 + (y_1-y_3)^2 + (x_2-x_3)^2 + (y_2-y_3)^2 \tag{5.4}$$

and the mean squared distance

$$MSD = TSD/3 \tag{5.5}$$

The probability that a cluster as tight as the active cluster would have arisen by chance involves the calculation of MSD for all the other possible clusters of three compounds. The number of clusters with an MSD value equal to or less than the active MSD is denoted by A (including the active cluster) and a probability is calculated as

$$p = A/N \tag{5.6}$$

where N is the total number of possible clusters of that size, in this case three compounds. It is obvious from inspection of the figure that there is one other cluster as tight as or tighter than the active cluster (compounds 2, 3, and 4) and that all other clusters have larger MSD values since they include compounds 1, 5, or 6. There are 20 possible clusters of three compounds in this set and thus $A = 2$, $N = 20$, and

$$p = 2/20 = 0.10 \tag{5.7}$$

If a probability level of 0.05 or less (95 per cent certainty or better) is taken as a significance level then this cluster of actives would be regarded as fortuitous.

Figure 5.12 shows a plot of a set of inhibitors of the enzyme monoamine oxidase (MAO) described by steric (E_s^c) and hydrophobic (π) parameters. It can be seen that the seven active compounds mostly cluster in the top left-hand quadrant of the plot. The original data set involved a dummy parameter, D, to indicate substitution by OCH_3 or OH at a particular position, and in the application of CSA to this problem, a set of random

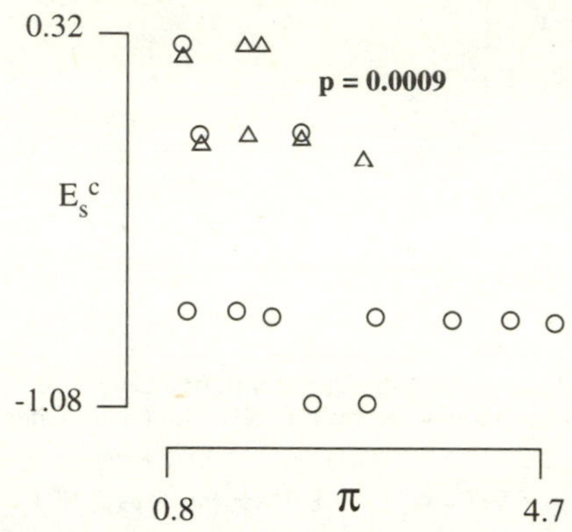

Fig. 5.12. Plot of active (Δ) and inactive (O) inhibitors of monoamine oxidase (from McFarland and Gans 1986, copyright (1986) American Chemical Society).

Table 5.10 Application of CSA to a set of 20 MAO inhibitors (from McFarland and Gans 1986 copyright (1986) American Chemical Society)

Parameters	A*	*p*
D	21464	0.27688
RN	14825	0.19124
πc	1956	0.02523
E_s^c	118	0.00152
D, π	1299	0.01676
D, E_s^c	1175	0.01516
RN, E_s^c	172	0.00222
π, E_s^c	71	0.00092
RN, π, E_s^c	151	0.00195
D, π, E_s^c	78	0.00101

* From a total possible set of 77,520 subsets of 7.

numbers, RN, was added to the data. The results of CSA analysis for this data are shown in Table 5.10 where it is seen that lowest probability of fortuitous clustering is given by the combination of π and E_s^c.

This illustrates another feature of CSA; not only can it be used to judge the significance of a particular set of clusters, it can also be used to test the effect (on the tightness of clusters) of adding or removing a particular descriptor. Thus, it may be used as a selection criterion for the usefulness of parameters. One thing that should be noted from the table is the large number of possible subsets (77,520) that can be generated for this data set. This may cause problems in the analysis of larger data sets in terms of the amount of computer time required. An approach to solving this problem is to compute a random sample of the possible combinations rather than exhaustively examining them all (McFarland and Gans 1986). CSA has been compared with three other QSAR techniques in the analysis of three different data sets (McFarland and Gans 1987).

5.6 Summary

Unsupervised learning methods, like the display techniques described in Chapter 4, are very useful in the preliminary stages of data analysis. Cluster analysis and FA produce easily understood displays from high-dimensional data sets and may be used when the number of variables in the set exceeds the number of samples. Although care must be exercised in the choice of class members when using *k*-nearest-neighbours, this and other methods described in this chapter should be reasonably safe from the danger of chance correlations. Cluster significance analysis allows us to attempt to assign significance levels to any 'interesting' groupings of samples seen using these methods or multivariate display techniques. Finally, in common with all of the other methods described in this book, it is not possible to say that any one technique is 'best'.

References

Chatfield, C. and Collins, A. J. (1980). *Introduction to multivariate analysis*. Chapman & Hall, London.

Chu, K. C., Feldman, R. J., Shapur, M. B., Hazard, G. F., and Geran, R. I. (1975). *Journal of Medicinal Chemistry*, **18**, 539–45.

Cormack, R. M. (1971). *Journal of the Royal Statistical Society*, **A134**, 321–67.

Dizy, M., Martin-Alvarez, P. J., Cabezudo, M. D., and Polo, M. C. (1992). *Journal of the Science of Food and Agriculture*, **60**, 47–53.

Ford, M. G., Greenwood, R., Turner, C. H., Hudson, B., and Livingstone, D. J. (1989). *Pesticide Science*, **27**, 305–26.

Goux, W. J. and Weber, D. S.(1993). *Carbohydrate Research*, **240**, 57–69.

Hudson, B. D., George, A. R., Ford, M. G., and Livingstone, D. J. (1992). *Journal of Computer-aided Molecular Design*, **6**, 191–201.

Jackson, J. E. (1991). *A user's guide to principal components*. Wiley, New York.

Kowalski, B. R. and Bender, C. F. (1972). *Analytical Chemistry*, **44**, 1405–11.

Lewi, P. J. (1976). *Arzneimettel-Forschung*, **26**, 1295–1300.

Li-Chan, E., Nakai, S., and Wood, D. E. (1987). *Journal of Food Science*, **52**, 31–41.

Malinowski, E. R. (1991). *Factor analysis in chemistry*. Wiley, New York.

McFarland, J. W. and Gans, D. J. (1986). *Journal of Medicinal Chemistry*, **29**, 505–14.

McFarland, J. W. and Gans, D. J. (1987). *Journal of Medicinal Chemistry*, **30**, 46–9.

Scarminio, I. S., Bruns, R. E., and Zagatto, E. A. G. (1982). *Energia Nuclear e Agricultura*, **4**, 99–111.

Svoboda, P., Pytela, O., and Vecera, M. (1983). *Collection Czechoslovak Chemical Communications*, **48**, 3287–306.

Takagi, T., Shindo, Y., Fujiwara, H., and Sasaki, Y. (1989). *Chemical and Pharmaceutical Bulletin*, **37**, 1556–60.

Willett, P. (1987). *Similarity and clustering in chemical information systems*. Research Studies Press, Wiley, Chichester.

6
Regression analysis

6.1 Introduction

Regression analysis is one of the most commonly used analytical methods in chemistry, including all of its specialist subdivisions and allied sciences. Indeed, the same can probably be said about most forms of science. The reason for its appeal lies perhaps in the fact that the method formalises something that the human pattern recogniser does instinctively, and that is to fit a line or a curve through a set of data points. We are accustomed to looking for trends in the data that the world presents to us, whether it be unemployment or inflation figures, or the results of some painstakingly performed experiments. We do this in the hope, or expectation, that the trends will reveal some underlying explanation of how or why the data is produced. In its simplest form, regression analysis involves fitting a straight line through a set of data points represented by just two variables, calculating an equation for the fitted line, and providing estimates of how well the points fit the line. The first section of this chapter will discuss simple linear regression and the calculation and interpretation of its statistics. The next section describes multiple linear regression: how the equations are constructed, non-linear regression models and the use of indicator variables in regression, including Free and Wilson analysis. The final section discusses some important features of regression analysis such as the comparison of regression models, tests for robustness, and the problems of chance correlations. Regression analysis based on variables derived from multivariate data, principal components, factors, and latent variables is discussed in Chapter 7, Supervised learning.

6.2 Simple linear regression

We have already seen in Chapter 1 an example of a simple linear regression model (eqn 1.2, Fig. 1.2) in which anaesthetic activity was related to the hydrophobicity parameter, π. How was the equation derived? If we consider the data shown plotted in Fig. 6.1, it is fairly obvious that a straight line can be fitted through the points. A line is shown on the figure and is described by the well-known equation for a straight line.

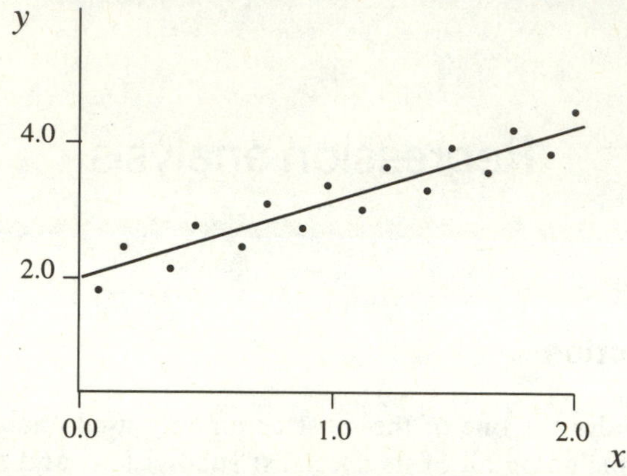

Fig. 6.1. Plot of the values of variable y against variable x with a fitted straight line.

$$y = mx + c \tag{6.1}$$

The value of c (2.0), the intercept of the line, can be read from the graph where $x = 0$ ($y = m0 + c$) and the value of m (1.0), the slope of the line, by taking the ratio of the differences in the y and x values at two points on the line $(y_2 - y_1)/(x_2 - x_1)$. A line such as this can be obtained easily by laying a straight edge along the data points and it is clear for this data that if another person repeated the procedure, a line with a very similar equation would result.

Figure 6.2 shows a different situation in which the data points still clearly correspond to a straight line but here it is possible to draw different lines through the data. Which of these two lines is best? Is it possible to say that one line is a better fit to the data than the other, or is some other line the best fit? Whether or not there is some way of saying what the 'right' answer is, it is clear that some objective way of fitting a line to data

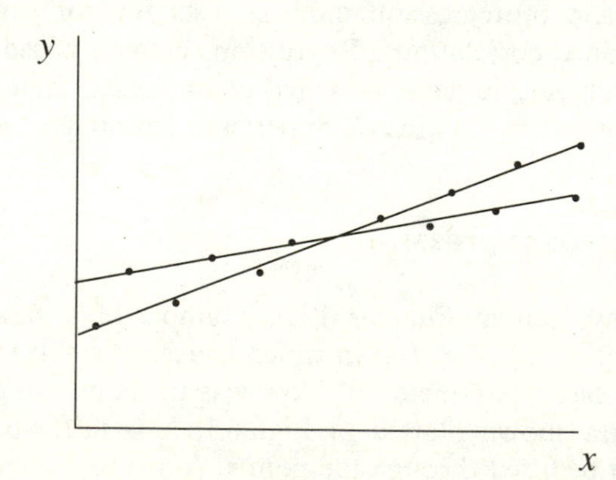

Fig. 6.2. Plot of y against x where two different straight lines can be fitted to the points.

such as those shown in these figures is required. One such technique is called the method of least squares, or ordinary least square (OLS), in which the squares of the distances between the points and the line are minimized. This is shown in Fig. 6.3 for the same data points as Fig. 6.2 with the exception that there is an extra data point in this figure. The extra point corresponds to the mean of the x ($\bar{x}$) and y ($\bar{y}$) data,

$$\bar{x} = \frac{\sum_{i=1}^{n} x_i}{n}, \quad \bar{y} = \frac{\sum_{i=1}^{n} y_i}{n} \tag{6.2}$$

and it can be seen that the regression line, or least-squares line, passes through this point. Since the point $(\bar{x}, \bar{y})$ lies on the line, the equation can be written as

$$y - \bar{y} = m(x - \bar{x}) \tag{6.3}$$

the constant term, c, having disappeared since it is explained by the means

$$c = \bar{y} - m\bar{x}. \tag{6.4}$$

Equation (6.3) can be rewritten as

$$y = \bar{y} + m(x - \bar{x}) \tag{6.5}$$

and thus for any value of x (x_i) an estimate of the y value ($\hat{y}_i$) can be made

$$\hat{y}_i = \bar{y} + m(x_i - \bar{x}). \tag{6.6}$$

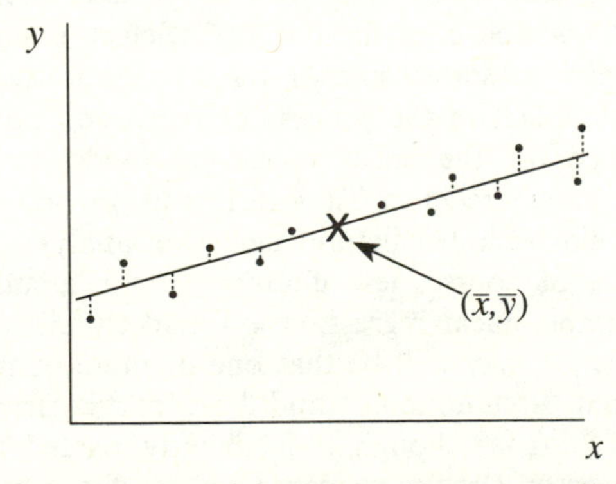

Fig. 6.3. Illustration of the process of least squares fitting.

The error in prediction for the y value corresponding to this x value is given by

$$y_i - \hat{y}_i = y_i - \bar{y} - m(x_i - \bar{x}). \tag{6.7}$$

This equation can be used to express a set of errors for the prediction of y values over the whole set of data (n pairs of points) and the sum of the squares of these errors is given by

$$U = \sum_{i=1}^{n} [y_i - \bar{y} - m(x_i - \bar{x})]^2. \tag{6.8}$$

Minimization of this sum of squares gives the slope of the regression line (m), which is equivalent to minimizing the lengths of the dotted lines, shown in Fig. 6.3, between the data points and the fitted line. It can be shown for minimum U (where $\mathrm{d}U/\mathrm{d}m = 0$ and $\mathrm{d}^2U/\mathrm{d}m^2$ is positive) that the slope is given by

$$m = \frac{\sum_{i=1}^{n} (x_i - \bar{x})(y_i - \bar{y})}{\sum_{i=1}^{n} (x_i - \bar{x})^2}. \tag{6.9}$$

Thus both the slope and the intercept of the least-squares line can be calculated from simple sums using eqns (6.4) and (6.9). In practice, few people ever calculate regression lines in this way as even quite simple scientific calculators have a least-squares fit built in. However, it is hoped that this brief section has illuminated the principles of the least-squares process and has shown some of what goes on in the 'black box' of regression packages.

Having fitted a least-squares line to a set of data points, the question may be asked, 'how well does the line fit?'. Before going on to consider this, it is necessary to state some of the assumptions, hitherto unmentioned, that are implicit in the process of regression analysis and which should be satisfied for the linear regression model to be valid. These assumptions are summarized in Table 6.1, and, in principle, all of these assumptions should be tested before regression analysis is applied to the data. In practice, of course, few if any of these assumptions are ever checked but if simple linear regression is found to fail when applied to a particular data set, it may well be that one or more of these assumptions have been violated. Assumptions 1 and 2 are particularly important since the data should look as though it is linearly related and at least the majority of the error should be contained in the y variable (called a

Table 6.1 Assumptions for simple linear regression

1	The x and y data is linearly related
2	The error is in y, the dependent variable
3	The average of the errors is 0
4	The errors are independent; there is no serial correlation among the errors (knowing the error for one observation gives no information about the others)
5	The errors are of approximately the same magnitude
6	The errors are approximately normally distributed (around a mean of zero)

regression of y on x). In many chemical applications this latter assumption will be quite safe as the dependent variable will often be some experimental quantity whereas the descriptor variables (the x set) will be calculated or measured with good precision.

The assumption of a normal distribution of the errors allows us to put confidence limits on the fit of the line to the data. This is carried out by the construction of an analysis of variance table (the basis of many statistical tests) in which a number of sums of squares are collected.* The total sum of squares (TSS), in other words the total variation in y, is given by summation of the difference between the observed y values and their mean.

$$\mathrm{TSS} = \sum_{i=1}^{n} (y_i - \bar{y})^2 \tag{6.10}$$

This sum of squares is made up from two components: the variance in y that is explained by the regression equation (known as the explained sum of squares, ESS), and the residual or unexplained sum of squares, RSS. The ESS is given by a comparison of the predicted y values ($\hat{y}$) with the mean

$$\mathrm{ESS} = \sum_{i=1}^{n} (\hat{y}_i - \bar{y})^2 \tag{6.11}$$

and the RSS by comparison of the actual y values with the predicted

$$\mathrm{RSS} = \sum_{i=1}^{n} (y_i - \hat{y}_i)^2. \tag{6.12}$$

The total sum of squares is equal to these two sums

---

* There is no agreed convention for abbreviating these sums of squares, other treatments may well use different sets of initials.

$$\text{TSS} = \text{ESS} + \text{RSS}. \tag{6.13}$$

These sums of squares are shown in the analysis of variance (ANOVA) table (Table 6.2). The mean squares are obtained by division of the sums of squares by the appropriate degrees of freedom. One degree of freedom is 'lost' with each parameter calculated from a set of data so the total sum of squares has $n-1$ degrees of freedom (where n is the number of data points) due to calculation of the mean. The residual sum of squares has $n-2$ degrees of freedom due to calculation of the mean and the slope of the line. The explained sum of squares has one degree of freedom corresponding to the slope of the regression line.

Table 6.2 ANOVA table

Source of variation	Sum of squares	Degrees of freedom	Mean square
Explained by regression	ESS	1	MSE (=ESS)
Residual	RSS	$n-2$	MSR (=RSS/$n-2$)
Total	TSS	$n-1$	MST (=TSS/$n-1$)

Knowledge of the mean squares and degrees of freedom allows assessment of the significance of a regression equation as described in the next section, but how can we assess how well the line fits the data? Perhaps the best known and most misused regression statistic is the correlation coefficient. The squared correlation coefficient (r^2) is given by division of the explained sum of squares by the total sum of squares

$$r^2 = \frac{\text{ESS}}{\text{TSS}}. \tag{6.14}$$

This can take a value of 0, where the regression is explaining none of the variance in the data, up to a value of 1 where the regression explains all of the variance in the set. r^2 multiplied by 100 gives the percentage of variance in the data set explained by the regression equation. The squared correlation coefficient is the square of the simple correlation coefficient, r, between y and x (see Panel in Chapter 2 p. 36). This correlation coefficient can take values between -1, a perfect negative correlation (y decreases as x increases), and $+1$, a perfect positive correlation. Correlation coefficients, both simple and multiple (where several variables are involved), can be very misleading. Consider the data shown in Fig. 6.4. Part (a) of the figure shows a set of data in which y is clearly dependent on x by a simple linear relationship; part (b) shows two separate 'clouds' of points where the line has been fitted between the two groups; parts (c) and (d) show two situations in which a single rogue point has greatly affected the fit of the

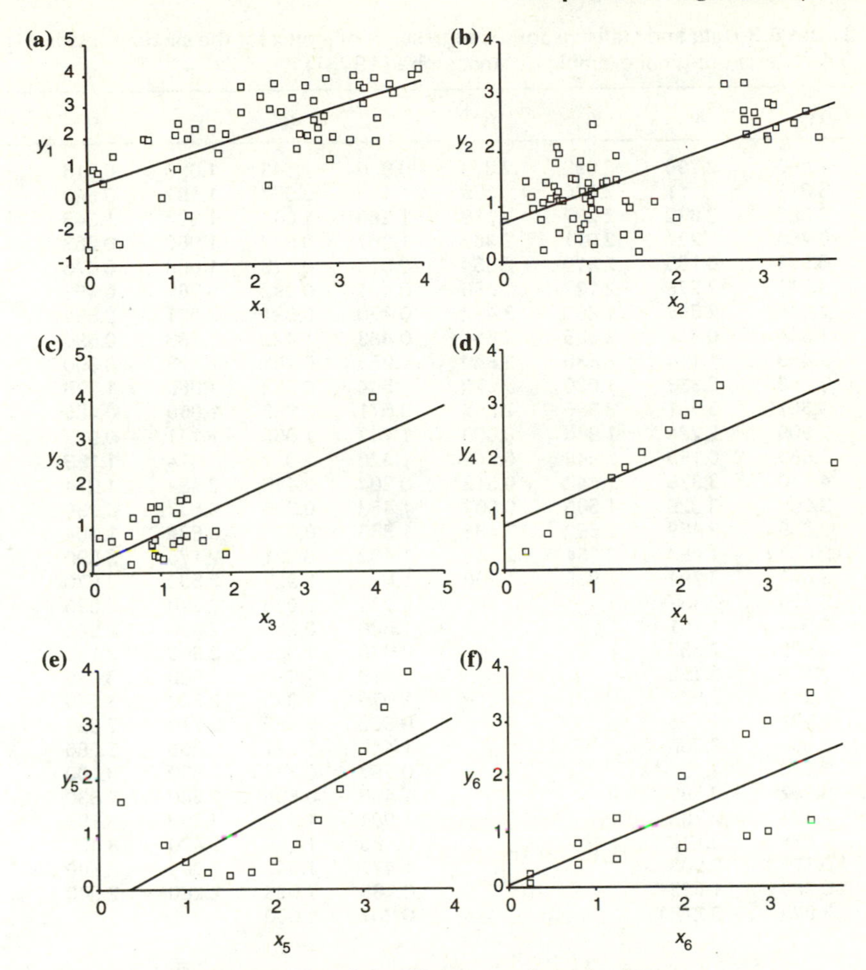

Fig. 6.4. Plot of six different sets of *y* and *x* data.

line.* Table 6.3 gives the data used to produce these plots and some of the statistics for the fit of the line. The correlation coefficients for these four graphs (and parts (e) and (f) of Fig. 6.4) are very similar, as are the regression coefficients for *x* (0.55 to 0.82). There is a somewhat wider range in values for the constant term (0.20 to 0.80), but overall the statistics give little indication of the four different situations shown in parts (a) to (d) of the figure. Parts (e) and (f) show two other types of data set for which a single straight line fit is inappropriate, a curve (e) and two separate straight lines (f). Once again the statistics give little warning, although the large

---

* This situation is often referred to as a 'point and cluster effect'; the regression line is fitted effectively between two points, the rogue point and the cluster of points making up the rest of the set.

Table 6.3 Data and statistics for the regression of y on x for the six data sets in Fig. 6.4 (after an original example by Anscombe (1973))*

y_1	x_1	y_1	x_1	y_2	x_2	y_2	x_2
1.898	2.790	2.092	2.573	0.920	1.149	1.090	0.716
3.318	2.111	2.820	1.879	1.138	0.577	1.187	0.648
3.385	3.672	2.287	1.318	1.266	1.047	1.902	1.343
−0.460	1.224	2.011	2.967	1.267	0.942	1.960	0.683
0.900	0.130	2.819	2.354	0.513	0.673	1.088	0.676
3.718	2.276	2.127	1.686	0.401	0.892	0.765	0.462
2.046	2.668	1.866	3.462	0.770	0.831	2.081	0.654
−1.344	0.374	3.655	2.644	0.483	1.422	1.783	0.637
2.459	1.114	3.630	1.880	1.258	0.661	1.178	0.350
3.559	3.333	1.020	0.070	0.594	0.922	1.065	1.779
3.667	3.631	3.865	2.897	0.671	0.962	1.069	0.485
2.909	2.224	1.986	0.689	1.702	1.055	1.141	0.977
0.589	0.198	1.948	0.749	1.420	1.017	1.414	1.188
4.150	3.973	1.442	0.312	0.204	0.483	2.484	1.074
3.066	3.326	−1.505	0.007	1.453	0.295	0.837	0.036
3.859	3.459	2.590	3.483	1.355	0.617	0.835	1.004
2.093	1.082	2.664	2.866	1.432	0.521	2.170	3.090
3.647	3.294	1.989	1.230	1.810	0.927	2.500	2.970
3.255	2.487			0.776	2.026	3.220	2.830
1.014	1.101			1.500	0.819	2.670	2.940
1.639	2.532			0.934	1.055	2.800	3.160
0.484	2.195			1.219	1.090	2.230	3.080
1.294	2.928			1.076	1.043	2.680	3.540
1.521	1.595			0.962	1.489	2.710	2.820
2.554	2.750			1.449	1.227	2.200	3.680
3.170	2.751			0.957	0.372	2.830	3.100
3.962	3.893			1.486	0.865	2.540	2.830
2.304	2.793			1.084	1.452	2.380	3.190
1.898	3.172			0.483	1.589	2.470	3.400
0.122	0.903			1.477	1.343	3.200	2.600
0.309	1.523			0.168	1.593	2.280	2.840
3.979	3.218			0.307	1.080		
		r=0.702				r=0.706	
		F=46.77				F=60.55	
		SE=0.95				SE=0.56	
		RC=0.82 (0.12)				RC=0.55 (0.07)	
		c=0.48 (0.12)				c=0.69 (0.12)	

* The statistics reported for each fit are—the simple correlation coefficient, r; the F statistic, F; the standard error of the fit, SE; the regression coefficient for x, RC, followed by its standard error in brackets; the constant of the equation, c, followed by its standard error in brackets.

standard errors of the constants do suggest that something is wrong. The lesson from this is clear; it is not possible to assess goodness of fit simply from a correlation coefficient. Indeed, this statistic (and others) can be very misleading, as well as very useful. In the case of a simple problem such as this involving just two variables the construction of a bivariate plot will reveal the patterns in the data. More complex situations present extra problems as discussed in the next section and Section 6.4.

Table 6.3—(continued)

y_3	x_3	y_4	x_4	y_5	x_5	y_6	x_6
0.191	0.542	0.333	0.250	1.600	0.250	0.250	0.250
1.270	0.575	0.665	0.500	0.813	0.750	0.800	0.800
1.536	0.961	0.998	0.750	0.500	1.000	1.250	1.250
0.943	1.772	1.663	1.250	0.313	1.250	2.000	2.000
0.742	0.895	1.862	1.400	0.250	1.500	2.750	2.750
0.837	1.361	2.128	1.600	0.313	1.750	3.000	3.000
0.851	0.517	2.527	1.900	0.500	2.000	3.500	3.500
0.399	0.900	2.793	2.100	0.813	2.250	0.080	0.250
1.369	1.215	2.993	2.250	1.250	2.500	0.400	0.800
0.643	0.854	3.325	2.500	1.813	2.750	0.510	1.250
0.740	1.590	1.900	3.800	2.500	3.000	0.700	2.000
0.405	1.914			3.313	3.250	0.910	2.750
0.731	0.275			3.950	3.500	0.995	3.000
0.733	1.262					1.195	3.500
0.807	0.095						
1.512	0.847						
1.233	0.903						
0.666	1.164						
0.619	0.850						
0.365	0.948						
1.692	1.365						
0.319	1.015						
1.667	1.268						
4.000	4.000						
r=0.702		r=0.709		r=0.707		r=0.708	
F=21.34		F=9.11		F=10.97		F=12.09	
SE=0.56		SE=0.72		SE=0.90		SE=0.80	
RC=0.72 (0.16)		RC=0.68 (0.22)		RC=0.85 (0.26)		RC=0.66 (0.19)	
c=0.20 (0.21)		c=0.80 (0.43)		c=−0.30 (0.57)		c=0.04 (0.42)	

6.3 Multiple linear regression

Multiple linear regression is an extension of simple linear regression by the inclusion of extra independent variables

$$y = ax_1 + bx_2 + \ldots + \text{constant}. \tag{6.15}$$

Least squares may be used to estimate the regression coefficients (a, b, c, and so on) for the independent variables (x_1, x_2, x_3, and so on), and the value of the constant term. Goodness of fit of the equation to the data can be obtained by calculation of a multiple correlation coefficient (R^2) just as for simple linear regression. In the case of simple linear regression it is easy to see what the fitting procedure is doing, i.e., fitting a line to the data, but what does multiple regression fitting do? The answer is that multiple

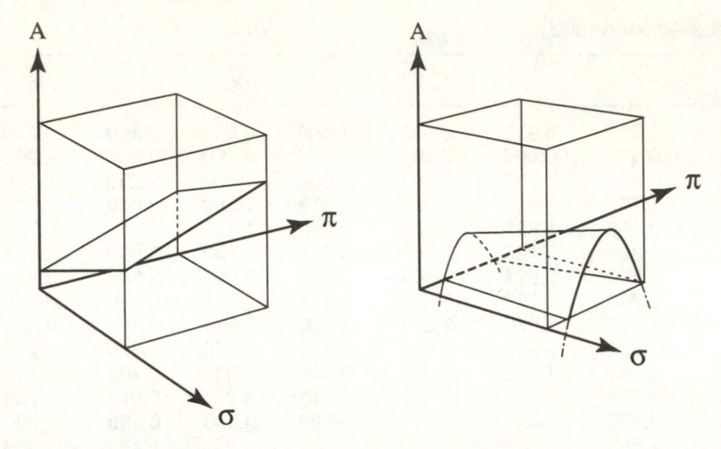

Fig. 6.5. Illustration of different surfaces corresponding to two-term regression equations.

regression fits a surface. Figure 6.5 shows the surface fitted by a two-term equation in π and σ (a plane) and an equation which includes a squared term. It is difficult to illustrate the results of fitting higher order equations but the principle is the same; multiple regression equations fit a surface to data with a dimensionality equal to the number of independent variables in the equation. It was shown in the previous section that the correlation coefficient can be a misleading statistic for simple linear regression fitting and the same is true for the multiple regression case. It is more difficult (or impossible) to check a multiple regression fit by plotting the data points with respect to all of the parameters in the equation, but one way that even the most complicated regression model can be evaluated is by plotting predicted y values against the observed values. If the regression equation is a perfect fit to the data ($R^2 = 1$), then a plot of the predicted versus observed should give a straight line with a slope of one and an intercept of zero. When some particular points are badly predicted it will be obvious from this plot; a curved plot suggests some other equation is more appropriate.

What about an assessment of the significance of the fit of a multiple regression equation (or simple regression) to a set of data? A guide to the overall significance of a regression model can be obtained by calculation of a quantity called the F statistic. This is simply the ratio of the explained mean square (MSE) to the residual mean square (MSR)

$$F = \frac{\text{MSE}}{\text{MSR}}. \tag{6.16}$$

An F statistic is used by looking up a standard value for F from a table of F statistics and comparing the calculated value with the tabulated value. If the calculated value is greater than the tabulated value, the equation is significant at that particular confidence level. F tables normally have values listed for different levels of significance, e.g., ten per cent, five per cent, and one per cent. As might be expected, the F values are greater for higher levels of significance. This is equivalent to saying that we expect the explained mean square to be even larger than the residual mean square in order to have a higher level of confidence in the fit. This seems like good common sense! Table 6.4 gives some values of the F statistic for different numbers of degrees of freedom at a significance level of five per cent. It can be seen that the table has entries for two degrees of freedom, the rows and the columns. These correspond to the number of degrees of freedom associated with the explained mean square, MSE, which is given by p (where p is the number of independent variables in the equation) and with the residual mean square, MSR, which is given by $n-p-1$ (where n is the number of data points). An F statistic is usually quoted as $F(\nu_1\ \nu_2)$, where $\nu_1=p$ and $\nu_2=n-p-1$. When regression equations are reported, it is not unusual to find the appropriate tabulated F value quoted for comparison with the calculated value.

The squared multiple correlation coefficient gives a measure of how well a regression model fits the data and the F statistic gives a measure of the overall significance of the fit.* What about the significance of individual terms? This can be assessed by calculation of the standard error of the regression coefficients, a measure of how much of the dependent variable

Table 6.4 Five per cent points (95 per cent confidence) of the F-distribution (reproduced from Lindley and Miller 1953, with permission of the Biometrika trustees)

	ν_1						
ν_2	1	2	3	4	5	10	∞
1	161.4	199.5	215.7	224.6	230.2	241.9	254.3
2	18.5	19.0	19.2	19.2	19.3	19.4	19.5
3	10.13	9.55	9.28	9.12	9.01	8.79	8.53
4	7.71	6.94	6.59	6.39	6.26	5.96	5.63
5	6.61	5.79	5.41	5.19	5.05	4.74	4.36
6	5.99	5.14	4.76	4.53	4.39	4.06	3.67
7	5.59	4.74	4.35	4.12	3.97	3.64	3.23
8	5.32	4.46	4.07	3.84	3.69	3.35	2.93
9	5.12	4.26	3.86	3.63	3.48	3.14	2.71
10	4.96	4.10	3.71	3.48	3.33	2.98	2.54
15	4.54	3.68	3.29	3.06	2.90	2.54	2.07
20	4.35	3.49	3.10	2.87	2.71	2.35	1.84
30	4.17	3.32	2.92	2.69	2.53	2.16	1.62
40	4.08	3.23	2.84	2.61	2.45	2.08	1.51
∞	3.84	3.00	2.60	2.37	2.21	1.83	1.00

* As long as the data is well distributed and does not behave as the examples shown in Fig. 6.4. This situation is often difficult to check for multivariate data.

prediction is contributed by that term. A statistic, the *t* statistic, may be calculated for each regression coefficient by division of the coefficient by its standard error (SE).

$$t = \left| \frac{b}{\text{SE of } b} \right| \tag{6.17}$$

Like the *F* statistic, the significance of *t* statistics is assessed by looking up a standard value in a table; the calculated value should exceed the tabulated value. Table 6.5 gives some values of the *t* statistic for different degrees of freedom and confidence levels. Unlike the *F* tables, *t* tables have only one degree of freedom which corresponds to the degree of freedom associated with the error sum of squares. This value is given by $(n-p-1)$, where n is the number of samples in the data set and p is the number of independent variables in the equation, including the constant. It can be seen from the table that the value of *t* at a five per cent significance level, for a reasonable number of degrees of freedom (five or more), is around two. This is equivalent to saying that the regression coefficient should be at least twice as big as its standard error if it is to be considered significant. Again, this seems like good common sense.

Another useful statistic that can be calculated to characterize the fit of a regression model to a set of data is the standard error of prediction. This gives a measure of how well one might expect to be able to make

Table 6.5 Percentage points of the *t*-distribution (reproduced from Lindley and Miller 1953, with permission of the Biometrika trustees)

	P						
ν_1	25	10	5	2	1	0.2	0.1
1	2.41	6.31	12.71	31.82	63.66	318.3	636.6
2	1.60	2.92	4.30	6.96	9.92	22.33	31.60
3	1.42	2.35	3.18	4.54	5.84	10.21	12.92
4	1.34	2.13	2.78	3.75	4.60	7.17	8.61
5	1.30	2.02	2.57	3.36	4.03	5.89	6.87
6	1.27	1.94	2.45	3.14	3.71	5.21	5.96
7	1.25	1.89	2.36	3.00	3.50	4.79	5.41
8	1.24	1.86	2.31	2.90	3.36	4.50	5.04
9	1.23	1.83	2.26	2.82	3.25	4.30	4.78
10	1.22	1.81	2.23	2.76	3.17	4.14	4.59
12	1.21	1.78	2.18	2.68	3.05	3.93	4.32
15	1.20	1.75	2.13	2.60	2.95	3.73	4.07
20	1.18	1.72	2.09	2.53	2.85	3.55	3.85
24	1.18	1.71	2.06	2.49	2.80	3.47	3.75
30	1.17	1.70	2.04	2.46	2.75	3.39	3.65
40	1.17	1.68	2.02	2.42	2.70	3.31	3.55
60	1.16	1.67	2.00	2.39	2.66	3.23	3.46
120	1.16	1.66	1.98	2.36	2.62	3.16	3.37
∞	1.15	1.64	1.96	2.33	2.58	3.09	3.29

individual predictions. In the situation where the standard error of measurement of the dependent variable is known, it is instructive to compare these two standard errors. If the standard error of prediction of the regression model is much smaller than the experimental standard error then the model has 'over-fitted' the data, whatever the other statistics of the fit might say. After all, it should not be possible to predict y with greater precision than it was measured, from a model derived from the experimental y values. Conversely, if the prediction standard error is much larger than the experimental standard error, then the model is unlikely to be very useful, although in this case it is likely that the other statistics will also indicate a poor fit. Where the experimental standard error is unknown the standard error of prediction can still be used to assess fit by comparison with the range of measured values. As a rule of thumb, if the prediction standard error is less than ten per cent of the range of measurements the model will be useful. For many data sets, particularly from biological experiments, a prediction within ten per cent may be regarded as very good. A summary of the statistics that have been described so far is shown in Table 6.6.

Table 6.6 Statistics used to characterize regression equations

Statistic		Use
Correlation coefficient	r	Gives the direction (sign) and degree (magnitude) of a correlation between two variables
Multiple correlation coefficient	R^2	A measure of how closely a regression model fits a data set
F statistic	F	A measure of the overall significance of a regression model
t statistic	t	A measure of the significance of individual terms in a regression equation
Standard error of prediction	SE	A measure of the precision with which predictions can be made from a regression equation

6.3.1 Creating multiple regression models

Creation of a simple linear regression equation is obvious; there are just two variables involved and all that is required is the estimation of the slope and intercept parameters, usually by OLS. The construction of multiple linear regression equations, on the other hand, is by no means as clear since the selection of independent variables for the equation involves choice. How can this choice be made? One obvious strategy is to use all of the independent variables and this in fact is the basis of one technique for the creation of multiple regression equations, backward-stepping regression. This procedure, as the name implies, begins by construction of a single linear regression model which contains all of the independent variables. Each term in this equation is examined for its contribution to the model, by comparison of t statistics, for example. The variable making the smallest contribution is removed and the regression model is recalculated, now with one term fewer. Any of the usual regression statistics can be used to assess the fit of this new model to the data and the procedure can be continued until a satisfactory multiple regression equation is obtained. Satisfactory here may mean an equation with a desired correlation coefficient or a particular number of physicochemical properties, etc.

Forward-stepping regression is—as might be expected—the reverse of this procedure. It begins with an equation containing only a single physicochemical parameter and involves adding terms one at a time. However, what may be surprising is that the application of a forward-stepping and backward-stepping procedure to the same data set does not necessarily yield the same answer. Newcomers to data analysis may find this disturbing and for some this may reinforce the prejudice that 'statistics will give you any answer that you want', which of course it can. The explanation of the fact that forward and backward stepping procedures can lead to different models lies in the presence of collinearity and multicollinearity in the data. A multiple regression equation may be viewed as a set of variables which *between* them account for some or all of the variation in the dependent variable. If the independent variables themselves are correlated in pairs (collinearity) or as linear combinations (multicollinearity) then different combinations may account for the same part of the variance of y. An example of this can be seen in eqns (6.18) and (6.19) which describe the pI_{50} for the inhibition of thiopurine methyltransferase by substituted benzoic acids in terms of calculated atomic charges (Kikuchi 1987).

$$pI_{50} = 12.5q_{2\pi} - 8.3 \tag{6.18}$$

$$n = 15 \quad r = 0.757$$

$$pI_{50} = 12.5q_{6\pi} - 8.4 \tag{6.19}$$

$$n = 15 \quad r = 0.785$$

The individual regression equations between pI_{50} and the two π-electron density parameters have quite reasonable correlation coefficients and thus it might be expected that they would be useful in a multiple regression equation. The two equations, however, are almost identical, indicating a very high collinearity between these descriptors. When combined into a two-term equation (eqn 6.20), which has an improved correlation coefficient, we see the effect of this collinearity; even the sign of one of the coefficients is changed.

$$pI_{50} = -74q_{2\pi} + 84q_{6\pi} - 6.3 \tag{6.20}$$

$$n = 15 \quad r = 0.855$$

The fact that these two descriptors are explaining a similar part of the variance in the pI_{50} values was revealed in the statistics of the fit for the two-term equation (high standard errors of the regression coefficients). Collinearity and multicollinearity in the descriptor set (independent variables) may lead to poor fit statistics or may cause instability in the regression coefficients. Indeed, regression coefficients which are seen to change markedly as variables are added to or removed from a model are a good indication of the presence of collinear variables. Are there solutions to these problems in the construction of multiple linear regression equations? Does forward- or backward-stepping regression give the best model? Are there alternative ways to construct multiple linear regression equations? Unfortunately, there are at least as many solutions as there are problems and there is also a large variety of procedures for the construction of regression models. One popular approach is to calculate all possible equations and then select the best on the basis of the fit statistics. The speed of modern computers allows such calculations to be carried out routinely but this procedure will be particularly prone to chance effects (see Section 6.4.2). The use of orthogonal variables, as described in the next chapter, can overcome the problems of collinearity but the question of variable selection for the models remains.

6.3.2 Non-linear regression models

Non-linear models may be fitted to data sets by the inclusion of functions of physicochemical parameters in a linear regression model—for example, an equation in π and π^2 as shown in Fig. 6.5—or by the use of non-linear fitting methods. The latter topic is outside the scope of this book but is well covered in many statistical texts (e.g. Draper and Smith 1981). Construction of linear regression models containing non-linear terms is most often prompted when the data is clearly not well fitted by a linear model, e.g. Fig. 6.4e, but where regularity in the data suggests that some other model will fit. A very common example in the field of quantitative structure–activity relationship (QSAR) involves non-linear relationships with hydrophobic descriptors such as log P or π. Non-linear dependency of biological properties on these parameters became apparent early in the development of QSAR models and a first approach to the solution of these problems involved fitting a parabola in log P (Hansch 1969).

$$\log 1/C = a(\log P)^2 + b \log P + c\sigma + \text{constant} \qquad (6.21)$$

Equation (6.21) may simply contain terms in π or log P, or may contain other parameters such as σ and so on. Unfortunately, many data sets appear to be well fitted by a parabola, as judged by the statistics of the fit, but in fact the data only corresponds to the first half or so of the curve. This is demonstrated in Fig. 6.6 for the fit shown in eqn (6.22).

$$\log 1/C = 1.37 \log P - 0.35(\log P)^2 + 2.32 \qquad (6.22)$$

Dissatisfaction with the fit of parabolic models such as this and a natural desire to 'explain' QSARs has led to the development of a number of mechanistic models, as discussed by Kubinyi (1993). These models give rise to various expected functional forms for the relationship between biological data and hydrophobicity, and data sets may be found which will be well fitted by them. Whatever the cause of such relationships it is clear that non-linear functions are required in order to model the biological data. An interesting feature of the use of non-linear functions is that it is possible to calculate an optimum value for the physicochemical property involved (usually log P). For example, eqn (6.22) gives an optimum value (at which log $1/C$ is a maximum) for log P of 1.96. This ability to derive optimum values led to attempts to define optima for the transport of compounds across various biological 'barriers'. For example, Hansch and co-workers (1968) examined a number of QSARs involving compounds acting on the central nervous system and concluded that the optimum log P value for penetration of the blood–brain barrier was 2. Subsequent work, by Hansch and others, has shown that the prediction of brain uptake is not quite such a simple matter. Van de Waterbeemd and Kansy (1992), for example, have

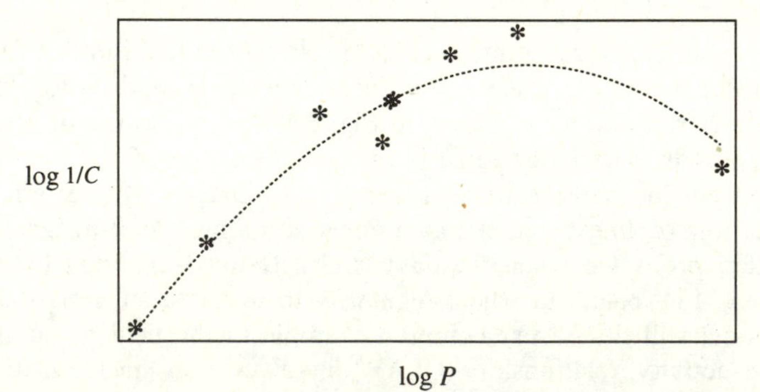

Fig. 6.6. Plot of biological response (C is the concentration to give a particular effect) against log P.

demonstrated that brain penetration may be described by a hydrogen-bonding capability parameter (Λ_{alk}) and Van der Waals' volume (V_m).

$$\log(C_{brain}/C_{blood}) = -0.338(\pm 0.03)\Lambda_{alk} + 0.007(\pm 0.001)V_m + 1.73(\pm 0.30)$$

$$n = 20 \quad r = 0.934 \quad s = 0.290 \quad F = 58 \qquad (6.23)$$

where the figures in brackets are the standard errors of the regression coefficients and s is the standard error of prediction.

6.3.3 Regression with indicator variables

Indicator variables are nominal descriptors (see Chapter 3) which can take one of a limited number of values, usually two. They are used to distinguish between different classes of members of a data set. This situation most commonly arises due to the presence or absence of specific chemical features; for example, an indicator variable might distinguish whether or not compounds contain a hydroxyl group, or have a *meta* substitution. An indicator variable may be used to combine two data sets which are based on different parent structures. Clearly, the dependent data for the different sets should be from the same source, otherwise there would be little point in combining them, and there should be some common physicochemical descriptors (but see later in this section, Free–Wilson method). Indicator variables are treated in multiple regression just as any other variable with regression coefficients computed by least squares. An example of this can be seen in the correlation of reverse phase HPLC capacity factors and calculated octanol/water partition coefficients for the xanthene and thioxanthene derivatives shown in Fig. 6.7 (Fillipatos *et al.* 1993). The correlation is given by eqn (6.24) in which the term D was used to indicate the presence ($D = 1$) or absence ($D = 0$) of the –NHCON(NO)– group, in other words series I or series II in Fig. 6.7.

$$\log P = 0.813(\pm 0.027)\log k_w + 2.114(\pm 0.161)D \qquad (6.24)$$

$$n = 24 \quad r = 0.972 \quad s = 0.365$$

Examination of the log k_w values showed that the replacement of oxygen by sulphur did not produce the expected increase in lipophilicity and it was found that a second indicator variable, S, to show the presence or absence of sulphur could be added to the equation to give

$$\log P = 0.768(\pm 0.021)\log k_w + 2.115(\pm 0.115)D + 0.415(\pm 0.095)S \quad (6.25)$$

$$n = 24 \quad r = 0.985 \quad s = 0.260$$

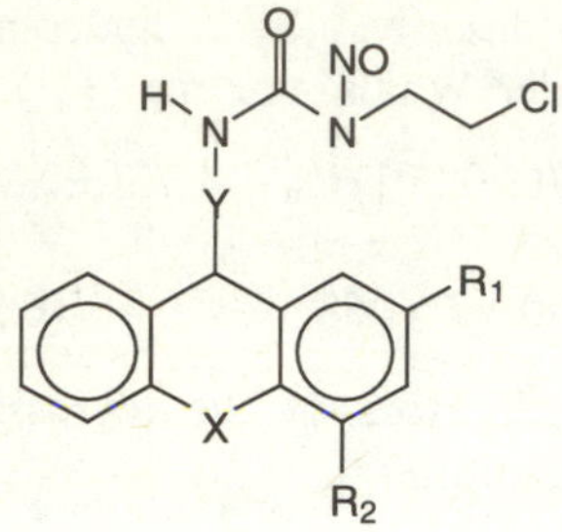

Fig. 6.7. Parent structures for the compounds described by eqn 6.24 and 6.25 (reproduced with permission from Fillipatos *et al.* 1993, copyright 1993 ESCOM Science Publishers B.V.).

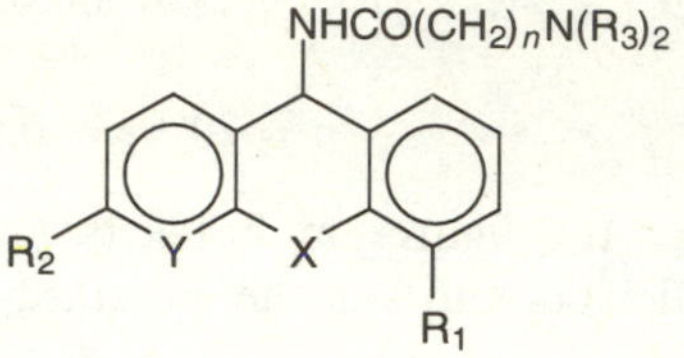

The correlation coefficient for eqn (6.25) is slightly improved over that for eqn (6.24) (but see section 6.4.3), the standard error has been reduced, and the regression coefficients for the log k_w and D terms are more or less the same. This demonstrates that this second indicator variable is explaining a different part of the variance in the log P values. It may have been noticed that eqns (6.24) and (6.25) do not contain intercept terms: this is because the intercepts are not significantly different to zero. These examples show how indicator variables can be used to improve the fit of regression models, but do the indicator variables (actually their regression coefficients) have any physicochemical meaning? The answer to this question is a rather unsatisfactory 'yes and no'. The sign of the regression coefficient of an indicator variable shows the direction (to reduce or enhance) of the effect of a particular chemical feature on the dependent variable while the size of the coefficient gives the magnitude of the effect. This does not necessarily bear any relationship to any particular physicochemical property, indeed it may be a mathematical artefact as described later. On the other hand, it may be possible to ascribe some meaning to indicator variable regression coefficients. The log P values used in eqns (6.24) and (6.25) were calculated by the Rekker fragmental method (see Section 9.2.1 and Table 9.2). This procedure relies on the use of fragment values for particular chemical groups and the –NHCON(NO)– group, accounted for by the indicator D, was missing from the scheme. The regression coefficient for this indicator variable has a fairly constant value, 2.114 in eqn (6.24) and 2.115 in eqn (6.25), suggesting that this might be a reasonable estimate for the fragment contribution of this group. Measurement of log P values for two compounds in set I allowed an estimate of $-2.09(\pm 0.14)$ to be made for this fragment, in good agreement with the regression coefficient of D. At first sight this statement may seem surprising since the signs of the fragment value and regression coefficients are different. The calculated log P values used in the equations did not take account of the hydrophilic (negative contribution) nitrosureido fragment and thus are bigger, by 2.11, than the experimentally determined HPLC capacity factors.

How does an indicator variable serve to merge two sets of data? The effect is difficult to visualize in multiple dimensions but can be seen in two dimensions in Fig. 6.8. Here, the two lines represent the fit of separate linear regression models, for multiple linear regression these would be surfaces. If the indicator variable has a value of zero for the compounds in set A it will have no effect on the regression line, whatever the value of the fitted regression coefficient. For the compounds in set B, however, the indicator variable has the effect of adding a constant to all the log $1/C$ values ($1 \times$ regression coefficient of the indicator variable). This results in a displacement of the regression line for the B subset of compounds so that it merges with the line for the A subset.

An indicator variable can be very useful in combining two subsets of compounds in this way since it allows the creation of a larger set which *may* lead to more reliable predictions. It is also useful to be able to describe the activity of compounds which are operating by a similar mechanism but which have some easily identified chemical differences. However, the situation portrayed in Fig. 6.8 is ideal in that the two regression lines are of identical slope and the indicator variable simply serves to displace them. If the lines were of different slopes the indicator may still merge them to produce an apparently good fit to the larger set, but in this case the fitted line would not correspond to a 'correct' fit for either of the two subsets. This situation is easy to see for a simple two-dimensional case but would clearly be difficult to identify for multiple linear regression. A way to ensure that an indicator variable is not producing a spurious, apparently good, fit is to model the two subsets separately and then compare these equations with the equation using the indicator. The situation can become even more complicated when two or more indicator

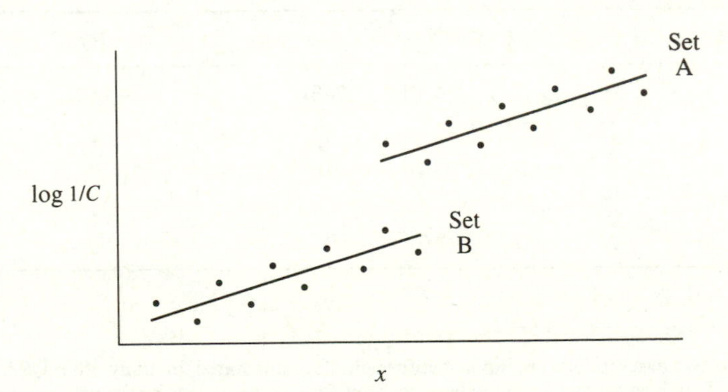

Fig. 6.8. Illustration of two subsets of compounds with different (parallel) fitted lines.

variables are used in multiple regression equations; great care should be taken in the interpretation of such models.

An interesting technique which dates from the early days of modern QSAR, known as the Free–Wilson method (Free and Wilson 1964) represents an extreme case of the use of indicator variables, since regression equations are generated which contain no physicochemical parameters. This technique relies on the following assumptions.

1. There is a constant contribution to activity from the parent structure.
2. Substituents on the parent make a constant contribution (positive or negative) to activity and this is additive.
3. There are no interaction effects between substituents, nor between substituents and the parent.

Of these assumptions, 1 is perhaps the most reasonable and 3 the most unreasonable. After all, it is the interaction of substituents with the electronic structure of the parent that gives rise to Hammett σ constants (see panel in Chapter 1, p. 6). However, despite any misgivings concerning the assumptions,* this method has the attractive feature that it is not necessary to measure or calculate any physicochemical properties; all that is required are measurements of some dependent variable. The technique operates by the generation of a data table consisting of zeros and ones. An example of such a data set is given in Table 6.7 for six compounds based on the parent structure shown in Fig. 6.9. A Free–Wilson table will also contain a column or columns of dependent (measured) data; for the example shown in Table 6.7 results were given for minimum inhibitory concentration (MIC) against two bacteria, *Mycobacterium tuberculosis* and *Mycobacterium kansasii*. Each column in a Free–Wilson data table, corresponding to a particular substituent at a particular position, is treated

Table 6.7 Free–Wilson data table (from Waisser *et al.* 1993 with kind permission)

Compound	R_1				R_2		
	H	4–CH_3	4–Cl	3–Br	H	4′–CH_3	4′–OCH_3
XI	1	0	0	0	1	0	0
XII	1	0	0	0	0	1	0
XIII	1	0	0	0	0	0	1
XIX	0	1	0	0	0	0	1
XXX	0	0	1	0	0	1	0
XXXV	0	0	0	1	1	0	0

* The first two assumptions are implicit, although often not stated, in many other QSAR/QSPR methods. The third assumption may be accounted for to some extent by the deliberate inclusion of several examples of each substituent.

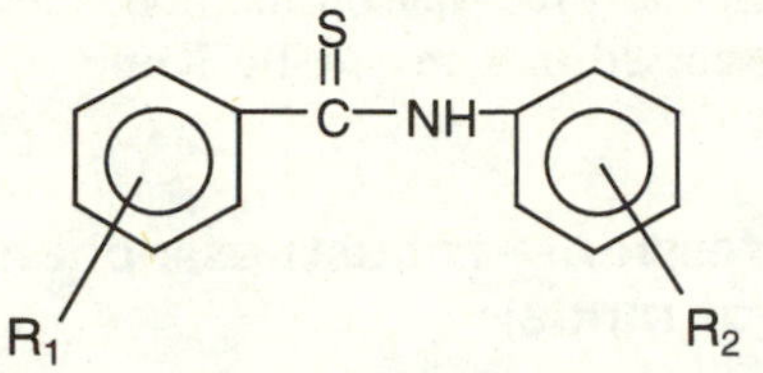

Fig. 6.9. Parent structure for the compounds given in Table 6.7 (from Waisser *et al.* 1993, with kind permission).

as an independent variable. A multiple regression equation is calculated in the usual way between the dependent variable and the independent variables with the regression statistics indicating goodness of fit. The regression coefficients for the independent variables represent the contribution to activity of that substituent at that position, as shown in Table 6.8.

One of the disadvantages of the Free–Wilson method is that—unlike regression equations based on physicochemical parameters—it cannot be used to make predictions for substituents not included in the original analysis. The technique may break down when there are linear dependencies between the structural descriptors, for example, when two substituents at two positions always occur together, or where interactions between substituents occur. Advantages of the technique include its ability to handle data sets with a small number of substituents at a large number of positions, a situation not well handled by other analytical methods, and its ability to describe quite unusual substituents since it does not require substituent constant data. A number of variations and improvements have

Table 6.8 Activity contributions for substituents as determined by the Free–Wilson technique (from Waisser *et al.* 1993 with kind permission)

Substituent	ΔMIC against	
	M. kansasii[a]	*M. tuberculosis*[b]
4–H	−0.397	−0.116
4–CH_3	0.264	0.101
4–OCH_3	0.290	0.337
4–Cl	0.095	−0.101
3–Br	−0.253	−0.312
4′–H	−0.078	0.088
4′–CH_3	0.260	0.303
4′–OCH_3	−0.081	0.085
4′–Cl	0.403	0.303
3′,4′–Cl_2	−0.259	−0.586
4′–C–C_6H_{11}	−0.589	−0.399
4′–Br	0.345	0.205
μ_o[c]	1.871	1.887

[a] Fit statistics, $r = 0.774$, $s = 0.43$, $F = 3.59$, $n = 35$.
[b] $r = 0.745$, $s = 0.42$, $F = 3.01$, $n = 35$.
[c] μ_o is the (constant) contribution of the parent structure to MIC.

been made to the original Free–Wilson method, these and applications of the technique are discussed in a review by Kubinyi (1988).

6.4 Multiple regression—robustness, chance effects, and the comparison of models

6.4.1 Robustness (cross-validation)

The preceding sections have shown how linear regression equations, both simple and multiple, may be fitted to data sets and statistics calculated to characterize their fit. It has also been shown how at least one statistic, the correlation coefficient, can give a misleading impression of how well a regression model fits a data set. This was shown in Fig. 6.4 which also demonstrates how easily this may be checked for a simple two-variable problem. A plot of predicted versus observed goes some way towards verification of the fit of multiple regression models but is there any other way that such a fit can be checked? One answer to this problem is a method known as cross-validation or jack-knifing. This involves leaving out a number of samples from the data set, calculating the regression model and then predicting values for the samples which were left out. Cross-validation is not restricted to the examination of regression models; it can be used for the evaluation of any method which makes predictions and, as will be seen in the next chapter, may be used for model selection.

How are the left-out samples chosen? One obvious way to choose these samples is to leave one out at a time (LOO) and this is probably the most commonly used form of cross-validation. Using the LOO method it is possible to calculate a cross-validated R^2, by comparison of predicted values (when the samples were not used to calculate the model) with the measured dependent variable values. This is also sometimes referred to as a prediction R^2. Such correlation coefficients will normally be lower than a 'regular' correlation coefficient and are said to be more representative of the performance (in terms of prediction) that can be expected from a regression. Other 'predictive' statistics, such as predicted residual sum of squares (PRESS, see Chapter 7), can also be calculated by this procedure. Cross-validation can not only give a measure of the likely performance of a regression model, it can also be used to assess how 'robust' or stable the model is. If the model is generally well fitted to a set of data then omission of one or more sample points should not greatly disturb the regression coefficients. By keeping track of these coefficients as samples are left out, it is possible to evaluate the model for stability, and also to identify which points most affect the fit.

Although LOO cross-validation is the most obvious choice, is it the best? Unfortunately, it is not. Figure 6.10 shows a simple two-dimensional

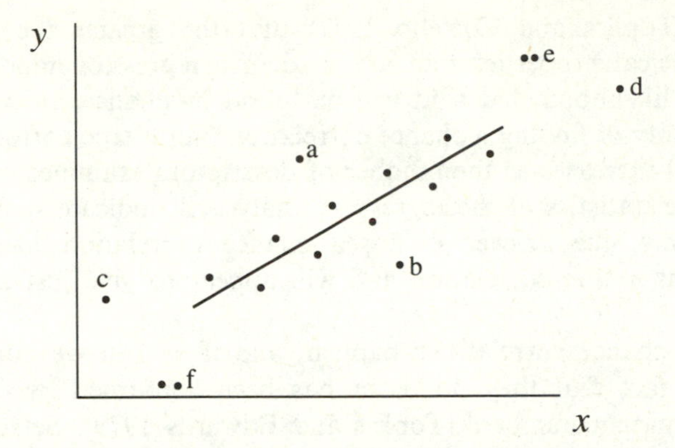

Fig. 6.10. Two-dimensional example of a data set with outliers.

situation in which a straight line model is well fitted to a set of data points which also contains a few outliers. Some of these points (a and b) will not affect the fitting of the line to the rest of the data and so will be badly predicted (whether included in the model or not) but would not alter the regression coefficients. Other points (c and d) which lie off the line but outside the rest of the data will affect the fit and thus will be badly predicted when left out and will alter the coefficients of the model. So far, so good—LOO cross-validation would identify these points. Samples e and f, however, occur along with another point well removed from the line and thus LOO would not identify them as being poorly predicted. A solution to this problem might be to leave compounds out in groups but the question then arises as to how to choose the groups. Cross-validation in groups can also result in the need for a lot of computer time to carry out the recalculation of the models and can generate a lot of information which needs to be assessed. Cross-validation is a useful technique for the assessment of fit and predictive performance of regression (and other) models but it is not the perfect measure that some have proposed it to be. A good solution to the questions of robustness and predictive performance is to use well-selected training and test sets, but this is a luxury we cannot always afford.

6.4.2 Chance effects

One of the problems with regression analysis, and other supervised learning methods, is that they seek to fit a model. This may seem like a curious statement to make, to criticize a method for doing just what it is intended to do. The reason that this is a problem is that given sufficient opportunity to fit a model then regression analysis will find an equation to fit a data set. What is meant by 'sufficient opportunity'? It has been proposed (Topliss and Costello 1972) that the greater the number of physicochemical properties that are tried in a regression model then the greater the likelihood that a fit will be found by chance. In other words, the probability of finding a chance correlation (not a true correlation but a coincidence) increases as the number of descriptors examined is increased. Will not the statistics of the regression analysis fit indicate such an effect? Unfortunately, the answer is no; a chance correlation has the same properties as a true correlation and will appear to give just as good (or bad) a fit.

Do such chance correlations happen, and if so can we guard against them? The fact that they do occur has been confirmed by experiments involving random numbers (Topliss and Edwards 1979). Sets of random numbers were generated, one set chosen as a dependent variable and several other sets as independent variables, and the dependent fitted to the independents using multiple regression. This procedure was repeated many times and occasionally a 'significant' correlation was found. A plot of average R^2 versus the number of random variables screened, for data sets containing different numbers of samples, is shown in Fig. 6.11. As originally proposed, the probability of finding a chance correlation for a given size of data set increases as the number of screened variables is increased. Plots such as that shown in the figure may be used to limit the number of variables examined in a regression study although it should be pointed out that these results apply to random numbers, and real data might be expected to behave differently. Perhaps the best test of the significance of a regression model is how well it performs with a real set of test data.

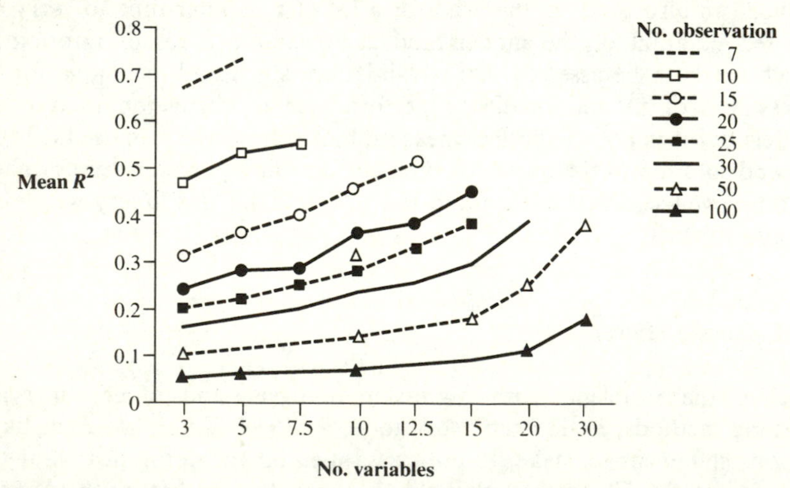

Fig. 6.11. Plot of mean R^2 versus the number of screened variables for regression equations generated for sets of random numbers (from Topliss and Edwards 1979, copyright (1979) American Chemical Society).

6.4.3 Comparison of regression models

If two regression equations contain the same number of terms and have been fitted to the same number of data points, then comparison is simple. The R^2 values will show which equation has the best fit and the F and t statistics may be used to judge the overall significance and the significance of individual terms. Obviously, if one equation is to be preferred over another it is expected that it will have significant fit statistics. Other factors may influence the choice of regression models, such as the availability or ease of calculation of the physicochemical descriptors involved. If the regression equations involve different numbers of terms (independent variables), then direct comparison of their correlation coefficients is not meaningful. Since the numerator for the expression defining the multiple correlation coefficient (eqn 6.14) is the explained sum of squares, it is to be expected that this will increase as extra terms are added to a regression model. Thus, R^2 would be expected to increase as extra terms are added. An alternative statistic to R^2, which takes account of the number of terms in a regression model, is known as the adjusted R^2 coefficient ($\bar{R}^2$)

$$\bar{R}^2 = 1 - (1 - R^2)\frac{n-1}{n-p} \tag{6.26}$$

where n is the number of data points and p the number of terms in the equation. This statistic should be used to compare regression equations with different numbers of terms. Finally, if two equations are fitted to different numbers of data points, selection of a model depends on how it is to be used.

6.5 Summary

Regression analysis is a very useful tool for the identification and exploitation of quantitative relationships. The fit of regression models may be readily estimated and the direction and magnitude of individual correlations may give some useful clues as to mechanism. Regression models are easily interpreted, since they mimic a natural process by which we try to relate cause and effect, but it should be remembered that a correlation does not prove such a relationship. Successful regression models may inspire us to design experiments to examine causal relationships and, of course, empirical predictions are always of use. There are dangers in the use of regression analysis—even quite simple models may be very misleading if judged by their statistics alone—but there are means by which some of the dangers may be guarded against. This chapter has been a brief introduction to some of the fundamentals of regression analysis; for further reading see Draper and Smith (1981), Montgomery and Peck (1982), or Rawlings (1988).

References

Anscombe, F. J. (1973). *American Statistician*, **27**, 17–21.

Draper, N. R. and Smith, H. (1981). *Applied regression analysis*. Wiley, New York.

Fillipatos, E., Tsantili-Kakoulidou, A., and Papadaki-Valiirake, A. (1993). In *Trends in QSAR and molecular modelling 92* (ed. C. G. Wermuth), pp. 367–9. ESCOM, Leiden.

Free, S. M. and Wilson, J. W. (1964). *Journal of Medicinal Chemistry*, **7**, 395–9.

Hansch, C. (1969). *Accounts of Chemical Research*, **2**, 232–9.

Hansch, C., Steward, A. R., Anderson, S. M., and Bentley, J. (1968). *Journal of Medicinal Chemistry*, **11**, 1–11.

Kikuchi, O. (1987). *Quantitative Structure–Activity Relationships*, **6**, 179–84 .

Kubinyi, H. (1988). *Quantitative Structure–Activity Relationships*, **7**, 121–33.

Kubinyi, H. (1993). *QSAR: Hansch analysis and related approaches*, Vol. 1 of *Methods and principles in medicinal chemistry*, (ed. R. Mannold, P. Krogsgaard-Larsen, H. Timmerman), pp. 68–77. VCH, Weinheim.

Lindley, D. V. and Miller, J. C. P. (1953). *Cambridge elementary statistical tables*. Cambridge University Press.

Montgomery, D. C. and Peck, E. A. (1982). *Introduction to linear regression analysis*. Wiley, New York.

Rawlings, J. O. (1988). *Applied regression analysis; a research tool*. Wadsworth & Brooks, California.

Topliss, J. G. and Costello, R. J. (1972). *Journal of Medicinal Chemistry*, **15**, 1066–8.

Topliss, J. G. and Edwards, R. P. (1979). *Journal of Medicinal Chemistry*, **22**, 1238–44.

Van de Waterbeemd, H. and Kansy, M. (1992). *Chimia*, **46**, 299–303.

Waisser, K., Kubicova, L., and Odlerova, Z. (1993). *Collection of Czechoslovak Chemical Communications*, **58**, 205–12.

7
Supervised learning

7.1 Introduction

The common feature underlying supervised learning methods is the use of the property of interest, the dependent variable, to build models and select variables. Regression analysis, which warranted a chapter of its own because of its widespread use, is a supervised learning technique. Supervised methods are subject to the danger of chance effects (as outlined in Section 6.4.2 for regression) which should be borne in mind when applying them. The dependent variable may be classified, as used in discriminant analysis described in the first section of this chapter, or continuous. Section 7.3 discusses variants of regression, which make use of linear combinations of the independent variables; Section 7.4 describes supervised learning procedures for feature selection.

7.2 Discriminant techniques

The first two parts of this section describe supervised learning methods which may be used for the analysis of classified data. One technique, discriminant analysis, is related to regression while the other, SIMCA, has similarities with principal component analysis (PCA). The final part of this section discusses some of the conditions which data should meet when analysed by discriminant techniques.

7.2.1 Discriminant analysis

Discriminant analysis, also known as the linear learning machine,* is intended for use with classified dependent data. The data may be measured on a nominal scale (yes/no, active/inactive, toxic/non-toxic) or an ordinal scale (1,2,3,4; active, medium, inactive) or may be derived from continuous data by some rule (such as 'low' if <10, 'high' if >10). The objective of

---

* *Linear* discriminant analysis is equivalent to the linear learning machine. There are also procedures for non-linear discriminant analysis (as there are for non-linear regression) but these will not be considered here.

regression analysis is to fit a line or a surface *through* a set of data points; discriminant analysis may be thought of as an orthogonal process to this in which a line or surface is fitted in between two classes of points in a data set. This is illustrated in Fig. 7.1 where the points represent compounds belonging to one of two classes, A or B, and the line represents a discriminating surface. It is confusing, perhaps, that the discriminant function itself (shown by the dotted line in Fig. 7.1) does run through the data points; the discriminant surface represents some critical value of the discriminant function, often zero. Projection of the sample points onto this discriminant function yields a value for each sample and classification is made by comparison of this value with the critical value for the function. In this simple two-dimensional example the discriminant function is a straight line, in the case of a set of samples described by N physicochemical properties the discriminant function would be an N-dimensional hypersurface. The discriminant function may be represented as

$$W = a_1x_1 + a_2x_2 + \ldots a_nx_n \tag{7.1}$$

or more succinctly as

$$W = \sum_{i=1}^{n} a_ix_i \tag{7.2}$$

where the x_i's are the independent variables used to describe the samples and the a_i's are fitted coefficients. These coefficients are known as the discriminant weights and may be rescaled to give discriminant loadings (Dillon and Goldstein 1984) which are the loadings of the variables onto the discriminant function, reminiscent of principal component loadings, and which are in fact the simple correlation of each variable with the discriminant function. Two things may be noticed from this equation and

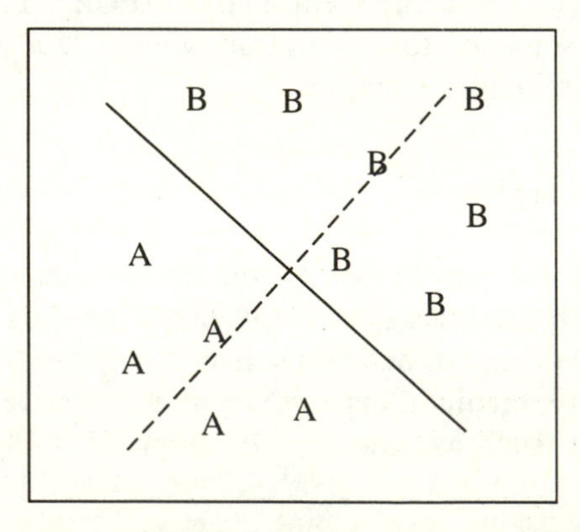

Fig. 7.1. Two-dimensional representation of discriminant analysis. The dotted line represents the discriminant function and the solid line a discriminant surface which separates the two classes of samples.

the figure. The combination of variables is a linear combination, thus this method strictly should be called linear discriminant analysis (LDA). The line shown drawn in the figure (which is at right angles to the discriminant function) is not the only line that could be drawn between the two classes of compounds. Creation of a different discriminant surface (drawing another line) is achieved by computing a different discriminant function. Unlike regression analysis, where the least squares estimate of regression coefficients can give only one answer, the coefficients of the variables in a discriminant function may take one of a number of values. As long as the discriminant function correctly classifies all of the samples, then it does not matter what the values of the discriminant loadings are. This is not to say that an individual discriminant analysis procedure run repeatedly on the same data set will give different answers, but that discriminant functions are more susceptible to change when samples or descriptors are added to or removed from the data set. The fact that a unique solution does not exist for a discriminant function has implications in the use of discriminant analysis for variable selection, as described in Section 7.4. Figure 7.1 also demonstrates how one of the alternative names, the linear learning machine, arose in artificial intelligence research. The algorithm to generate the discriminant function is a 'learning machine' which aims to separate the two classes of samples in a linear fashion by 'learning' from the data.

Returning to eqn (7.1), how is this used? Once a discriminant function has been generated, a prediction can be made for a compound by multiplication of the descriptor variables by their coefficients and summation of these products. This will give a value of W which, if it exceeds some critical value, will assign the compound to one of the two classes. A value of zero may be used as the critical value so that if W is positive the compound belongs to class 1 (A in Fig. 7.1), if negative then to class 2 (B in Fig. 7.1). The question also arises of how to judge the quality of a discriminant function. In terms of prediction, this is quite easily done by comparison of the predicted class membership with known class membership. A more stringent test of the predictive ability of discriminant analysis is to use a leave one out (LOO) cross-validation procedure as described for regression analysis in Section 6.4.1. It is also possible to compute a statistic for discriminant analysis which is equivalent to the F statistic used to characterize the fit of regression models. This statistic may be used to judge the overall significance of a discriminant analysis result and, in a slightly different form as a partial F statistic, also used to construct discriminant functions (see Dillon and Goldstein 1984).

An early example of the use of discriminant analysis in QSAR involved inhibition of the enzyme monoamine oxidase (MAO) by derivatives of aminotetralins and aminoindans shown in Fig. 7.2 (Martin *et al.* 1974, see also Section 5.5). These compounds inhibited the enzyme *in vitro*, and it

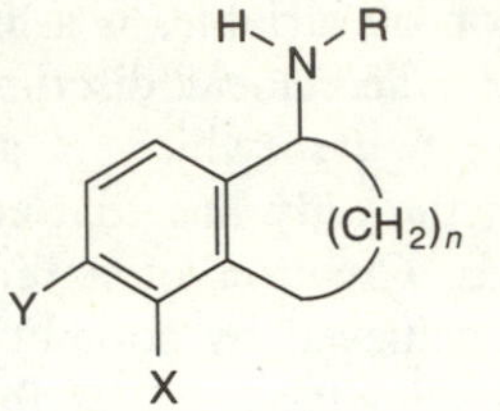

Fig. 7.2. Parent structure for the compounds shown in Table 7.1 (from Martin *et al.* 1974, copyright (1974) American Chemical Society).

was possible to obtain percentage inhibition data for them in an enzyme assay, but the crucial test was a measure of their activity *in vivo*. This was assessed by judgement of the severity of symptoms following administration of *dl*-Dopa and was given a score of 0, 1, 2, or 3 as shown in Table 7.1. Initial examination of the data by discriminant analysis suggested that the compounds should be classified into just two groups,* and thus a two-group discriminant function was fitted as shown in the table. This function involved the steric parameter, E_s^c, and an indicator variable which showed whether the compounds were substituted in position X or Y in Fig. 7.2. The results of classification by this function are quite impressive: only one compound (number 18) is misclassified. A more exacting test of the utility of discriminant analysis was carried out by fitting a discriminant function to 11 of the compounds in Table 7.1 (indicated by a *). Once again, E_s^c and the indicator variable were found to be important; one of the training set compounds was misclassified and all of the test set compounds were correctly assigned.

A comparison of the performance of discriminant analysis and other analytical techniques on the characterization of two species of ants by gas chromatography was reported by Brill and co-workers (1985). Samples of two species of fire ants, *Solenopsis invicta* and *S. richteri*, were prepared using a dynamic headspace analysis procedure. The gas chromatography analysis resulted in the generation of 52 features, retention data for cuticular hydrocarbons, for each of the samples. Each of these descriptors was examined for its ability to discriminate between the two species of ants and the three best features selected for use by the different analytical methods. A non-linear map of the samples described by these three features is shown in Fig. 7.3 where it is clear that the two species of ants are well separated. Analysis of this data was carried out by *k*-nearest-neighbour (see Section 5.2), discriminant analysis and SIMCA (see Section 7.2.2). The discriminant analysis routine used was the linear learning machine (LLM) procedure in the pattern recognition package ARTHUR (see Software appendix). A

---

* This was supported by the pharmacology since on retest, some compounds moved between groups 0 and 1 or 2 and 3, but rarely from (0,1) to (2,3) or vice versa.

Table 7.1 Structure, physicochemical properties, and potency as MAO inhibitors (*in vivo*) of the aminotetralins and aminoindans shown in Fig. 7.2 (from Martin *et al.* 1974, copyright (1974) American Chemical Society)

	Structure				Properties[a]		Potency		
Number	n[b]	R	X	Y	Π	E_s^c	Observed[c]	Calculated[d]	Calculated[e]
1	2	CH_3	H	OCH_3	1.3	0.00	3	1	1*
2	3	H	OCH_3	H	1.2	0.32	3	1	0*
3	3	H	H	OCH_3	1.3	0.32	3	1	1
4	3	CH_2CH_3	H	OCH_3	2.2	−0.07	3	1	1
5	3	CH_3	H	OCH_3	1.7	0.00	2	1	1*
6	3	CH_3	H	OH	1.7	0.00	2	1	1
7	2	H	H	OCH_3	0.8	0.32	2	1	1*
8	3	CH_3	OCH_3	H	1.7	0.00	1	0	0*
9	3	$(CH_2)_2OCH_3$	H	OCH_3	1.7	−0.66	1	0	0*
10	3	$(CH_2)_2CH_3$	H	OCH_3	2.7	−0.66	1	0	0*
11	3	$(CH_2)_5CH_3$	H	OCH_3	4.2	−0.68	1	0	0
12	3	$CH_2C_6H_5$	OCH_3	H	3.5	−0.68	1	0	0
13	3	$(CH_2)_2OH$	H	OCH_3	1.0	−0.66	1	0	0
14	3	CH_3	OH	H	1.7	0.00	0	0	0*
15	3	$CH(CH_3)_2$	OCH_3	H	2.6	−1.08	0	0	0
16	3	$CH(CH_3)_2$	H	OCH_3	2.6	−1.08	0	0	0*
17	2	$CH(CH_3)_2$	H	OCH_3	2.1	−1.08	0	0	0
18	2	H	OCH_3	H	0.8	0.32	0	1	0*
19	3	$(CH_2)CH_3$	H	OCH_3	1.4	−0.66	0	0	0*
20	3	$(CH_2)_6CH_3$	H	OCH_3	4.7	−0.68	0	0	0

[a] Two indicator variables were also used to distinguish indans and tetralins and the position of substitution (X or Y).
[b] Where n is the n in Fig. 7.2.
[c] Activity was scored (0–3) according to the severity of symptoms; scores 2 and 3 are active compounds, 0 and 1 inactive.
[d] Calculated from a two-group discriminant function.
[e] Calculated from a two-group discriminant function which was trained on half of the compound set (indicated by *).

training/test set protocol was used in which the data was split up into two sets five times, with all of the original data used as a test set member at least once. Results of these analyses are shown in Table 7.2. These rather impressive results are perhaps not surprising given the clear separation of the two species shown in Fig. 7.3. They do, however, illustrate an important general feature of data analysis and that is that there is no 'right' way to analyse a particular data set. As long as the method used is suitable for the data, in this case classified dependent data, then comparable results should be obtained if the data contains appropriate information.

These two examples have involved classified dependent data and continuous independent data. It is also possible to use classified independent data in discriminant analysis, as it is in regression (such as Free–Wilson analysis) or a combination of classified and continuous independent variables. Zalewski (1992) has reported a discriminant analysis of sweet and non-sweet cyclohexene aldoximes using indicator variables describing chemical structural features. From the discriminant function it was

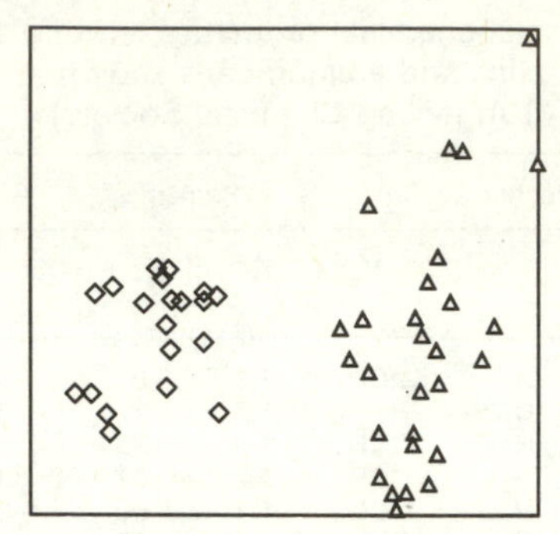

Fig. 7.3. Non-linear map of two species of fire ant, *Solenopsis invicta* (◇) and *Solenopsis. richteri* (△), described by three GC peaks (from Brill *et al.* 1985, with permission of Elsevier Science).

possible to identify features associated with the two classes of compounds:

Sweet	a short chain up to three carbons
	a substituted carbon at a particular position
	a cyclohexane ring
Non-sweet	a carbon chain at a different position to the sweet compounds
	the presence of a heteroatom in a ring

This function classified correctly 22 out of 23 sweet compounds and 24 out of 29 non-sweet derivatives. The same report also described a discriminant analysis of another set of aldoxime derivatives characterized by molecular connectivity indices (see Box 1.4 in Chapter 1 p. 16). The discriminant function involved just two of these indices (Kier 1980)

$$D_F = 1.21\,{}^1\chi - 3.88\,{}^4\chi_p \qquad (7.3)$$

The first term (${}^1\chi$) in eqn (7.3) describes the size of molecules in terms of the number of bonds, the second term (${}^4\chi_p$) is influenced by the size of the substituents. Compounds with a value of D_F greater than -3.27 were

Table 7.2 Classification of fire ants characterized by gas chromatography data (from Brill *et al.* 1985, with permission of Elsevier Science)

Method	Category*	Training correct (%)	Test correct (%)	Overall (%)
KNN	1	100	100	100
(*k*=10)	2	100	100	100
LLM	1	100	89.6	97.9
	2	100	100	100
SIMCA	1	100	100	100
	2	100	100	100

* 1=*S. richteri*, 2=*S. invicta*.

classified as sweet. This discriminant function correctly assigned nine out of ten sweet compounds and eight out of ten non-sweet.

How are discriminant functions built? The construction of a multiple variable discriminant function presents similar problems to the construction of multiple regression models (see Section 6.3.1) and similar solutions have been adopted. The stepwise construction of discriminant functions presents an extra problem in that there is not necessarily a unique solution, unlike the situation for regression. The other question that may have occurred to a reader by now is whether discriminant analysis is able to handle more than two classes of samples. The answer is yes, although the number of discriminant functions needed is not known in advance. If the classes are organized in a uniform way in the multidimensional space, then it may be possible to classify them using a single discriminant function, e.g., if W (from eqn 7.1) $< x$ then class 1, if $x < W < y$ then class 2, if $W > y$ then class 3 (where x and y are some numerical limits). When the classes are not organized in such a convenient way, then it will be necessary to calculate extra discriminant functions; in general, k classes will require $k-1$ discriminant functions.

An example of multi-category discriminant analysis can be seen in a report of the use of four different tumour markers to distinguish between controls (healthy blood donors) and patients with one of three different types of cancer (Lanteri *et al.* 1988). The results of this analysis are shown in Table 7.3 where LDA is compared with k-nearest-neighbour (KNN, k was not specified).

7.2.2 SIMCA

The SIMCA* method is based on the construction of principal components (PCs) which effectively fit a box (or hyper-box in N dimensions) around each class of samples in a data set. This is an interesting application of PCA, an unsupervised learning method, to different classes of data resulting in a supervised learning technique. The relationship between SIMCA and PLS, another supervized principal components method, can be seen clearly by reference to Section 7.3.

How does SIMCA work? The steps involved in a SIMCA analysis are as follows

1. The data matrix is split up into subsets corresponding to each class and PCA is carried out on the subsets.

---

* The meaning of these initials is variously ascribed to Soft Independent Modelling of Class Analogy, or Statistical Isolinear MultiCategory Analysis, or SImple Modelling of Class Analogy.

Table 7.3 Classification of cancer patients based on tumour marker measurements (from Lanteri *et al.* 1988, with permisson of John Wiley)

Discrimination between	LDA correct (%)[a]		KNN (%)
	Classification	Prediction	
Controls—patients[b]	83.1	83.7	85.7
Controls—lungs	92.8	91.2	86.8
Controls—breast	79.1	81.1	81.5
Controls—gastro	93.4	91.9	89.1

[a] The data from 102 subjects (30 controls, 72 patients) was divided into training set (classification) and test set (prediction) containing about 70 per cent and 30 per cent of the subjects respectively.
[b] The difference between controls and all patients.

2. The number of PCs necessary to model each class is determined (this effectively defines the hyper-box for each category).
3. The original descriptors are examined for their discriminatory power and modelling power (see below) and irrelevant ones discarded.
4. PCA is again carried out on the reduced data matrix and steps 2 and 3 repeated. This procedure continues until consistent models are achieved.

The discriminatory power and modelling power parameters are measures of how well a particular physicochemical descriptor contributes to the PCs in terms of the separation of classes and the position of samples within the classes. Since the PCs are recalculated when descriptors with low discriminatory or modelling power are removed, new parameters must be recalculated for the remaining properties in the set. Thus, the whole SIMCA analysis becomes an iterative procedure to obtain an optimum

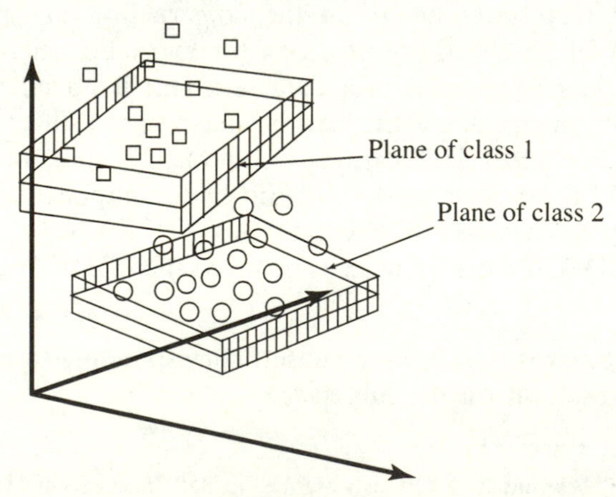

Fig. 7.4. Graphical representation of SIMCA (after Dunn *et al.* 1978, copyright (1978) American Chemical Society).

solution. Readers interested in further mathematical details of the SIMCA method should consult the chapter by Wold and Sjostrom (1977).

The results of applying the SIMCA procedure can perhaps best be seen in a diagram such as that shown in Fig. 7.4. The hyper-boxes do not fit around all of the points in each class but for the purposes of prediction it is possible to assign a sample to the nearest hyper-box. The size and shape of the hyper-boxes allows a probability of class membership to be assigned to predictions and if the objects within a class have some associated continuous property, it is possible to make quantitative predictions (by the position of a sample within the hyper-box compared to other points). The SIMCA technique has been applied to a variety of problems within the QSAR field and others. One data set that has already been cited (Section 5.2) was also analysed by SIMCA (Scarminio *et al.* 1982). This data consisted of water samples characterized by their concentrations of four elements (the four most important for classification were chosen from a total of 18 elements measured). A comparison of the performance of SIMCA and the worst nearest neighbour analysis is shown in Table 7.4. For a test set of seven samples, four Lindoya and three Serra Negra, the 9-NN analysis classified all correctly while the SIMCA method misclassified two of the Serra Negra samples as Valinhos.

Table 7.4 Comparison of SIMCA and KNN for classification of water samples (from Scarminio *et al.* 1982, with kind permission)

Region	Number of samples	Number of points incorrectly classified	
		9-NN	SIMCA
Serra Negra	46	2	16
Lindoya	24	2	5
São Jorge	7	1	0
Valinhos	39	2	3
Correct (%)		93.8	79.3

Another example of the use of SIMCA for the analysis of multicategory data was reported by Page (1986). This involved the characterization of orange juice samples of different varieties from various geographical locations using the following techniques: HPLC (34), inductively coupled plasma emission spectrometry (10), infrared (40), fluorescence (6), ^{13}C NMR (49), ultraviolet (29), enzymatic analysis (6), and GC (13). The numbers in brackets give the number of parameters measured using each method; the use of such a variety of analytical techniques allowed the identification of different components of the juices, e.g., trace elements, sugars, organic acids, polyphenols etc. The performance of SIMCA on the

Table 7.5 SIMCA classification of orange juice samples (from Page 1986, with permission of the Institute of Food Technologists)

True class[a]	Computed class—assigned (%)							
	1	2	3	4	5	6	7	8
1	100							
2		100						
3			82.4			11.8		5.9
4				94.4	5.6			
5				6.7	93.3			
6				16.7		83.3		
7							100	
8				18.2				81.8

[a] The classes are: 1, Valencia (Florida); 2, Valencia (other US); 3, Valencia (non-US); 4, Hamlin (all); 5, Pineapple (all); 6, Navel (all); 7, others (Florida); 8, others (Brazil).

HPLC data (carotenoids, flavones, and flavonoids) is shown in Table 7.5.* In this example SIMCA can be seen to have done quite an impressive job in classification; the lowest figure on the diagonal of the table is 81.8 per cent, and in most cases, wrong assignment is only made to one other class.

7.2.3 Conditions and cautions for discriminant analysis

As was said at the beginning of this chapter, supervised learning techniques in general are subject to the dangers of chance effects, and discriminant techniques are no exception. Jurs and co-workers have reported quite extensive investigations of the problem of chance separation (Stuper and Jurs 1976, Whalen-Pedersen and Jurs 1979, Stouch and Jurs 1985a and b). As was the case for regression analysis using random numbers, it was found that the probability of achieving a separation by chance using LDA increased as the number of variables examined was increased. The situation, however, for discriminant analysis is compounded by the question of the dimensionality of the data set. In the limit where a data set contains as many physicochemical (or structural, e.g., indicators) variables as there are samples in the set, there is a trivial solution to the discriminant problem. The discriminant procedure has as many adjustable parameters (the discriminant function coefficients) as there are data points and thus can achieve a perfect fit. This is equivalent to fitting a multiple linear regression model to a data set with zero degrees of freedom. The recommendation from Jurs's work is that the ratio of data points (compounds, samples, etc.) to descriptor variables should be three or greater.

* Interestingly, this type of table is known as a 'confusion matrix' (Dillon and Goldstein 1984).

Another aspect of the dimensionality of a data set that is perhaps not quite so obvious concerns the number of members in each class. Ideally, each of the classes in a data set should contain about the same number of members. If one class contains only a small number of samples, say ten per cent of the total points or less, then the discriminant function may be able to achieve a trivial separation despite the fact that the ratio of points to descriptors for the overall data set is greater than three. The following guidelines should be borne in mind when applying discriminant techniques.

- The number of variables employed should be kept to a minimum (by preselection) and the ratio $N:P$ (samples:parameters) should be greater than three.
- The number of members in each class should be about equal, if necessary by changing the classification scheme or by selecting samples.

Finally, it may be the case that the data is not capable of linear separation. Such a situation is shown in Fig. 7.5 where one class is embedded within the other. A recently described technique for the treatment of such data sets makes use of PCs scaled according to the parameter values of the class of most interest, usually the 'actives' (Rose *et al.* 1991). This is somewhat reminiscent of the SIMCA method.

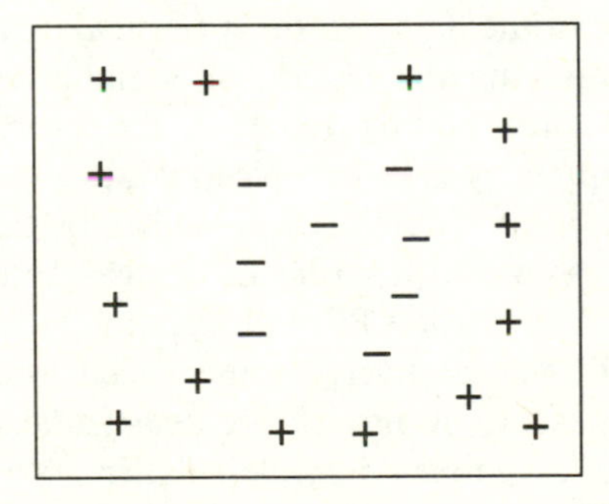

Fig. 7.5. Illustration of a data set in which one class is embedded in another.

7.3 Regression on principal components and partial least squares

Methods such as PCA (see Section 4.2) and factor analysis (FA) (see Section 5.3) are data-reduction techniques which result in the creation of new variables from linear combinations of the original variables. These new variables have an important quality, orthogonality, which makes them particularly suitable for use in the construction of regression models. They are also sorted in order of importance, in so far as the amount of variance in the independent variable set that they explain, which makes the choice of them for regression equations somewhat easier than the choice of 'regular' variables. Unfortunately, the fact that they are linear combinations makes the subsequent interpretation of regression models somewhat more difficult. The following sections describe regression using PCs, a variant of this called **partial least squares (PLS)**, and, briefly, a technique called continuum regression which embraces both of these and ordinary multiple regression.

7.3.1 Regression on principal components

The first step in carrying out **principal component regression (PCR)** is, unsurprisingly, a PCA. This produces scores and loadings as described by eqn (4.1), reproduced below.

$$\begin{aligned} PC_1 &= a_{1,1}v_1 + a_{1,2}v_2 + \ldots a_{1,N}v_N \\ PC_2 &= a_{2,1}v_1 + a_{2,2}v_2 + \ldots a_{2,N}v_N \\ PC_q &= a_{q,1}v_1 + a_{q,2}v_2 + \ldots a_{q,N}v_N \end{aligned} \tag{7.4}$$

The scores are the values of each PC for each compound, the loadings are the subscripted coefficients ($a_{i,j}$) in eqn (7.4). A score for a particular compound or sample in the data set (for a particular principal component) is computed by multiplying the descriptor variable values by the appropriate loadings and then adding together the products. Each PC has associated with it a quantity called an eigenvalue, a measure of how much of the variance in the original data set is described by that component. Since the components are calculated in order of decreasing amounts of variance explained, it follows that the first PC will have the largest eigenvalue in the set and subsequent PCs successively smaller eigenvalues. Can these eigenvalues be used as a measure of importance or 'significance'? After all, PCA will produce as many components as the smaller of P points or N dimensions,* so there may be as many PCs as there were originally dimensions. The answer to this question is a reassuring perhaps! If the original data is autoscaled (see Section 3.4), then each variable will contribute one unit of variance to the data set, which will have a total variance of N where N is the number of variables in the set. As PCs are produced, their eigenvalues will decrease until they fall below a value of one. At this point the components will no longer be explaining as much variance as one of the original variables in the set and this might make a reasonable limit to assess com-

---

* Actually, it is the rank of the matrix, denoted by $r(\mathbf{A})$, which is the maximum number of linearly independent rows (or columns) in **A**. $0 \leqslant r(\mathbf{A}) \leqslant \min(p,n)$, where A has p rows and n columns.

ponents as meaningful (however, see later). For most real data sets, components with an eigenvalue of approximately one are found in a far smaller number than the original number of properties.

Having carried out PCA, what comes next? The principal component scores are treated as any other variables would be in a multiple regression analysis and MLR models are constructed as shown in eqn (7.4)

$$y = a_1\mathrm{PC}_{x_1} + a_2\mathrm{PC}_{x_2} + \ldots a_n\mathrm{PC}_{x_n} + c \tag{7.5}$$

where y is some dependent variable, perhaps log $1/C$ for a set of biological results, a_1 to a_n are a set of regression coefficients fitted by least squares to the n principal components in the model and c is a constant (intercept). The fit of the regression model can be evaluated by using the usual regression statistics and the equation can be built up by forward-stepping, backward-stepping, or whatever (see Chapter 6). Is it possible to say which PCs should be incorporated into a PCR model? Surely the components are calculated in order of their importance and thus we might expect them to enter in the order one, two, three, and so on. This is partly true, components are calculated in order of their importance in terms of explaining the variance in the set of *independent* variables and very often the first one or two components will also be best correlated with a dependent variable. But for this to happen depends on a good choice of variables in the first place, in so far as they are correlated with y, and the fact that a linear combination of them will correlate with the dependent variable.

An example may illustrate this. In an attempt to describe the experimentally determined formation constants for charge-transfer complexes of monosubstituted benzenes with trinitrobenzene, a set of computational chemistry parameters were calculated (Livingstone *et al.* 1992). The initial data set contained 58 computed physicochemical descriptors, which after the removal of correlated variables (see Section 3.5), left a set of 31. Parameters were selected from this set, on the basis of their ability to describe the formation constants (see next section), to leave a reduced subset of 11 descriptors. PCA carried out on this set gave rise to four components with an eigenvalue greater than one, and one component with an eigenvalue close to one (0.95), as shown in Table 7.6. Forward-stepping regression analysis between κ, a substituent constant derived from the formation constants, and these PCs led to the following equations

$$\kappa = 0.191\mathrm{PC}_1 + 0.453 \tag{7.6}$$

$$R^2 = 0.5 \quad F = 33.01 \quad \mathrm{SE} = 0.32$$

$$\kappa = 0.191\mathrm{PC}_1 + 0.193\mathrm{PC}_4 + 0.453 \tag{7.7}$$

$$R^2 = 0.732 \quad F = 43.77 \quad \mathrm{SE} = 0.24$$

Table 7.6 Variable loadings* for the first five PCs derived from the reduced data set of 11 variables (from Livingstone *et al.* 1992, with permission of the Royal Society of Chemistry)

	Component (eigenvalue)				
	1 (2.73)	2 (2.19)	3 (1.78)	4 (1.23)	5 (0.95)
Variable			Loading		
CMR	0.48	−0.34			
C log *P*		−0.41		−0.47	
E_{HOMO}		−0.36		0.49	
P3	0.41				0.33
μ_x		0.48		−0.37	
Sn(1)	−0.31				0.42
Sn(2)			−0.59		
P1				−0.41	0.60
Fe(4)	−0.39			−0.38	
μ	−0.40	0.40			0.38
Sn(3)			−0.60		

* For simplicity, only loadings above 0.3 are shown.

$$\kappa = 0.191\mathrm{PC}_1 + 0.193\mathrm{PC}_4 + 0.130\mathrm{PC}_5 + 0.453 \qquad (7.8)$$

$$R^2 = 0.814 \quad F = 45.22 \quad \mathrm{SE} = 0.20$$

$n = 35$ for all three equations, the F statistics are all significant (at 99 per cent probability) and the t statistics for the individual regression coefficients are significant at greater than the one per cent level.

A number of things may be seen from these equations. The first component to be incorporated in the regression models was indeed the first PC and this, combined with a constant, accounted for half of the variance in the dependent variable set ($R^2 = 0.5$). The next component to enter the model, however, was PC_4 despite the fact that the variables in the set of 11 had been chosen for their *individual* ability to describe κ. Clearly, the linear combinations imposed by PCA, combined with the requirements of orthogonality, did not produce new variables in PC_2 and PC_3 which were useful in the description of κ. The third PC to be included has an eigenvalue of less than one and yet it is seen to be significant in eqn (7.8). If the eigenvalue cut-off of less than one had been imposed on this data set, eqn (7.8) would not have been found.

As extra terms are added to the regression model it can be seen that the regression coefficients do not change, unlike the case for MLR with untransformed variables where collinearity and multicollinearity amongst the descriptors can lead to instability in the regression coefficients. The regression coefficients in eqns (7.6) to (7.8) remain constant because the principal component scores are orthogonal to one another, the inclusion of extra terms in a PCR leads to the explanation of variance in the dependent variable not covered by terms already in the model. These features of PCR make it an attractive technique for the analysis of data; an unsupervised learning method (lower probability of chance effects?) produces a reduced set of well-behaved (orthogonal) descriptors followed by the production of a stable, easily interpreted, regression model. Unfortunately, although the regression equation is easily understood and applied for prediction, the relationship with the original variables is much more obscure. This is the big disadvantage of PCR, and related techniques such as PLS described in the next section. Inspection of Table 7.6 allows one to begin to ascribe some chemical 'meaning' to the PCs, for example, bulk factors (CMR and P3) load onto PC_1 and C log $P(-0.47)$ loads onto PC_4, but it should be remembered that the PCs are mathematical constructs designed to explain the variance in the x set. The use of varimax rotation (see Section 4.2) may lead to some simplification in the interpretation of PCs.

7.3.2 Partial least squares

The method of partial least squares (PLS) is also a regression technique which makes use of quantities like PCs derived from the set of independent variables. The PCs in PLS regression models are called latent variables (LV),* as shown in the PLS equation, eqn (7.9).

$$y = a_1\mathrm{LV}_1 + a_2\mathrm{LV}_2 + \dots a_n\mathrm{LV}_n \tag{7.9}$$

where y is a dependent variable and a_1 to a_n are regression coefficients fitted by the PLS procedure. Each latent variable is a linear combination of the independent variable set.

$$\mathrm{LV}_1 = b_{1,1}x_1 + b_{1,2}x_2 + \dots b_{1,n}x_n$$

$$\mathrm{LV}_2 = b_{2,1}x_1 + b_{2,2}x_2 + \dots b_{2,n}x_n$$

$$\mathrm{LV}_q = b_{q,1}x_1 + b_{q,2}x_2 + \dots b_{q,n}x_n \tag{7.10}$$

As in PCA, PLS will generate as many latent variables (q) as the smaller of N (dimensions) or P (samples). Thus far, PLS appears to generate identical models to PCR so what is the difference (other than terminology)? The answer is that the PLS procedure calculates the latent variables and the regression coefficients in eqn (7.9) all at the same time. The algorithm is actually an iterative procedure (Wold 1978) but the effect is to combine the

* This term may be generally used to describe variables derived from measured variables (for example, PCs, factors, etc.).

PCA step of PCR with the regression step. Latent variables, like PCs, are calculated to explain most of the variance in the x set while remaining orthogonal to one another. Thus, the first latent variable (LV_1) will explain most of the variance in the independent set, LV_2 the next largest amount of variance and so on. The important difference between PLS and PCR is that the latent variables are constructed so as to maximize their correlation with the dependent variable. Unlike PCR equations where the PCs do not enter in any particular order (see eqns 7.6 to 7.8) the latent variables will enter PLS equations in the order one, two, three, etc. The properties of latent variables are:

(1) the first latent variable explains maximum variance in the independent set; successive latent variables explain successively smaller amounts of variance.
(2) the latent variables conform to 1 with the provision that they are maximally correlated with the dependent variable.
(3) the latent variables are orthogonal to one another.

One problem with the PLS procedure, common to both PCR and multiple linear regression (MLR), is the choice of the number of latent variables to include in the model. The statistics of the fit can be used to judge the number of variables to include in an MLR but the situation is somewhat more complex for PCR and PLS. Judgement has to be exercised as to how 'significant' the LVs or PCs are. Although the statistics of the fit may indicate that a particular PC or LV is making a significant contribution to a regression equation, that variable may contain very little 'information'. The eigenvalue of a PC or LV may be a guide but as was seen in the previous section, some cut-off value for the eigenvalue is not necessarily a good measure of significance. A commonly used procedure in the construction of PLS models is to use leave one out (LOO) cross-validation (see Section 6.4.1) to estimate prediction errors. This works by fitting a PLS model to $n-1$ samples and making a prediction of the y value for the omitted sample ($\hat{y}$). When this has been carried out for every sample in the data set a predictive sum of squares (PRESS) can be calculated for that model.

$$\mathrm{PRESS} = \sum_{i=1}^{n} (y_i - \hat{y}_i)^2 \tag{7.11}$$

Note that this sum of squares looks similar to the residual sum of squares (RSS) given by eqn (6.12) but is different; in eqn (6.12) the $\hat{y}_i$ is predicted from an equation that includes that data point; here the $\hat{y}_i$ is not in the model hence the term *predictive* residual sum of squares. The difference in predictive ability of two PLS models can be evaluated by comparison of their PRESS values.

$$E = \frac{\text{PRESS}_i}{\text{PRESS}_{i-1}} \tag{7.12}$$

The E statistic compares a PLS model of i components with the model containing one component less and, in order to evaluate PRESS for the one component model, PRESS for the model containing no components is calculated by comparing predicted values with the mean. A critical value of 0.4 has been suggested for E (Wold 1978) and when this is exceeded, the PLS equation with i components is doing no better (or worse) in prediction than the model with $i-1$ latent variables.

On the face of it, PLS appears to offer a much superior approach to the construction of linear regression models than MLR or PCR (since the dependent variable is used to construct the latent variables) and for some data sets this is certainly true. Application of PLS to the charge-transfer data set described in the last section resulted in a PLS model containing only two dimensions which explained over 90 per cent of the variance in the substituent constant data. This compares very favourably with the two- and three-dimensional PCR equations (eqns 7.7 and 7.8) which explain 73 and 81 per cent of the variance respectively. Another advantage that is claimed for the PLS approach is its ability to handle redundant information in the independent variables. Since the latent variables are constructed so as to correlate with the dependent variable, redundancy in the form of collinearity and multicollinearity in the descriptor set should not interfere. This is demonstrated by fitting PLS models to the 31 variable and 11 variable parameter sets for the charge-transfer data. As shown in Table 7.7 the resulting PLS models account for very similar amounts of variance in κ.

How are PLS models used? One obvious way is to simply make predictions for test set samples by calculation of their latent variables from eqn (7.10) and application of the appropriate regression coefficients (eqn 7.9). The latent variables may be used like PCs for data display (see Chapter 4) by the construction of scores plots for samples and loadings plots for variables. A PLS analysis of halogenated ether anaesthetics allowed the production of the scores plot shown in Fig. 7.6 in which anaesthetics with similar side-effects are grouped together (Hellberg *et al.*

Table 7.7 Modelling κ by PLS (from Livingstone *et al.* 1992, with permission of the Royal Society of Chemistry)

	Percentage of κ variance explained using:	
PLS model of dimension:	11 variable dataset	31 variable dataset
1	78.6	78.7
2	92.9	90.4
3	94.9	95.1

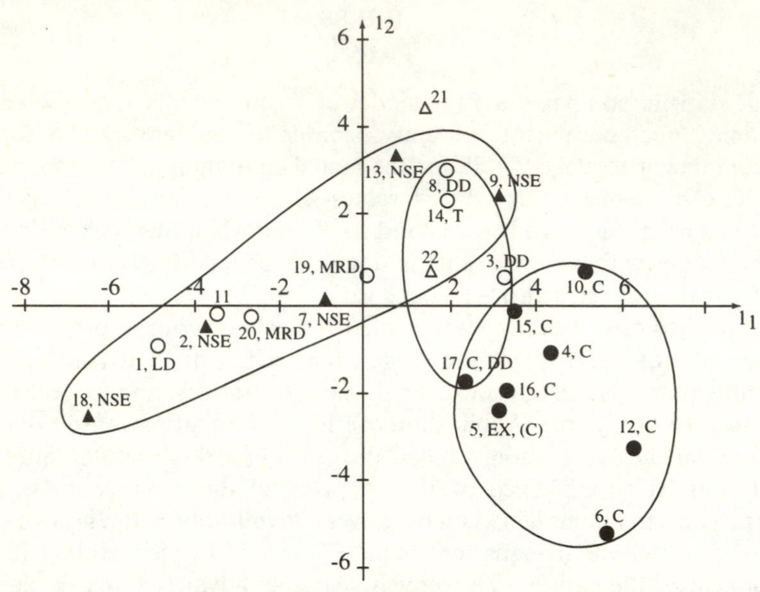

Fig. 7.6. A PLS scores plot for a set of halogenated ether anaesthetics; the ellipses enclose compounds with similar side-effects (from Hellberg *et al.* 1985, with permission of VCH, Weinheim).

1985). The biological data available for these compounds included a measure of toxicity and separate PLS models were fitted to the anaesthetic and toxicity data. Prediction of toxicity was good from a three-component PLS model as shown in Fig. 7.7.

Another widespread use of the PLS technique is based on its ability to handle very large numbers of physicochemical parameters. The increasing use of molecular modelling packages in the analysis of biological and other data has led to the establishment of so-called three-dimensional QSAR (Goodford 1985, Cramer *et al.* 1988, Kubinyi 1993). In these approaches a grid of points is superimposed on each of the molecules in the training set. Probes are positioned at each of the points on the grid and an interaction energy calculated between the probe and the molecule. Depending on the resolution chosen for the grid, several thousand energies may be calculated for each type of probe for every molecule in the set. Clearly, many of these energies will be zero or very small and may be discarded, and many will be highly correlated with one another. For example, when a positively charged probe is placed at the grid points near a region of high electron density, the attractive interactions will be similar. PLS is used to model the relationship between these grid point interaction energies and the dependent variable. The resultant PLS models may be visualized by displaying the 'important' grid points, as determined by their loadings onto the latent variables and the coefficients of these variables in the PLS regression equation.

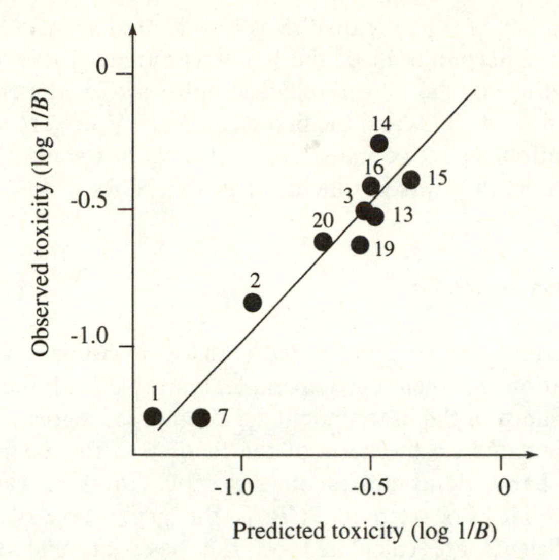

Fig. 7.7. Plot of toxicity predictions from a three-component PLS model (from Hellberg *et al.* 1985, with permission of VCH, Weinheim).

What are the problems with the PLS technique? One problem, like the question of deciding dimensionality, is shared with MLR and PCR. Why fit a linear model? The imposition of simple linear relationships on nature might be thought of as 'expecting too much'. Fortunately, simple or at least fairly simple linear relationships often do hold and linear models do quite a

Table 7.8 PC and LV* loadings for charge-transfer data (from Livingstone *et al.* 1992, with permission of the Royal Society of Chemistry)

	PC component (eigenvalue)		PLS latent variable	
	1 (2.73)	2 (2.19)	1	2
Variable	Loading		Loadings	
CMR	0.48	−0.34	0.48	
C log *P*		−0.41	−0.32	0.67
E_{HOMO}		−0.36		−0.51
P3	0.41		0.42	
μ_x		0.48		0.36
$S_n(1)$	−0.31		−0.24	
$S_n(2)$				
P1				
$F_e(4)$	−0.39		−0.34	
μ	−0.40	0.40	−0.39	0.4
$S_n(3)$				

* As in Table 7.6, only loadings above 0.3 are shown for clarity (except the loading for $S_n(1)$ on latent variable 1 so that it can be compared with PC_1).

good job. Another problem with PLS is also shared with PCR and that is the question of interpretation of the latent variables. Table 7.8 shows the important loadings of the 11 variable descriptor set for the charge-transfer data onto the first two PCs and the first two LVs. LV_1 is very similar to PC_1 with the exception that it contains C log *P*, LV_2 has some similarity with PC_2 but of course PC_2 was not included in the PCRs.

7.3.3 Continuum regression

In MLR the equations are constructed so as to maximize the explanation of the correlation between the dependent variable and the independent variables. Variance in the independent set is ignored, regression coefficients are simply calculated on the basis of the fit of y to the x variables. PCR, on the other hand, concentrates on the explanation of variance in the descriptor set. The first step in PCR is the generation of the PCs, regression coefficients are calculated on the basis of explanation of the correlation between y and these components.

These two approaches to the construction of regression models between y and an x set can be viewed as extremes. The relative balance between explanation of variance (in the x set) and correlation (y with x) can be expressed as a parameter, α, which takes the value of 0 for MLR and 1 for PCR. PLS regression sets out to describe both variance and correlation, and thus will have a value of α of 0.5, midway between these two extremes. Continuum regression is a new type of regression procedure which contains an adjustable parameter, α, which allows the production of all three types of regression model (Stone and Brooks 1990). An alternative formulation of continuum regression has been developed in which the parameter α is optimized during the production of the regression model (Malpass *et al.* 1994). The two most popular forms of regression analysis, MLR and PLS, tend to be applied to data sets in a quite arbitrary way, often dictated by the whim (or experience) of the analyst. Continuum regression presents the opportunity to allow the structure within a data set to determine the most appropriate value of α, and hence the decision as to which regression model to fit. Indeed, since α is an adjustable parameter which can adopt any value between zero and one, it is possible to fit models which do not correspond to MLR, PLS, or PCR. This is illustrated in Table 7.9 for some literature data sets where it can be seen that α values of 0.24, 0.35, 0.73, and 0.85 are obtained for some components of the fitted models. The first two of these correspond to models which are somewhere between MLR and PLS, while the latter two correspond to models in between PLS and PCR. The two- and three-dimensional models for the Wilson and Famini example have α values near zero which correspond to the MLR used in the original analysis. The example data set from Clark and co-workers, on the other hand, which was also originally fitted by MLR gives an α value of 0.55 corresponding to a PLS model. Continuum regression appears to offer the useful facility of allowing the data to select the model to fit.

Table 7.9 Continuum regression results for literature data

Reference[a]	Number of samples	Number of variables	R^2	α^b
Dunn *et al.* 1984 (PLS)	7	11	0.69 (1)[c]	0.73
Wilson and Famini 1991 (MLR)	21	6	0.94 (3)	0.35, 0.07, 0.07
Clark *et al.* 1986 (MLR)	25	3	0.85 (1)	0.55
Li *et al.* 1982 (MLR)	40	4	0.95 (3)	0.58, 0.85, 0.58
Diana *et al.* 1987 (MLR)	12	3	0.85 (2)	0.4, 0.24

[a] The method used in the original report is shown in brackets.
[b] Where more than one component is used, the α values are for models of each dimensionality.
[c] The number of components used for this R^2 is given in brackets.

7.4 Feature selection

One of the problems involved in the analysis of any data set is the identification of important features. So far this book has been involved with the independent variables, but the problem of feature selection may also apply to a set of dependent variables (see Chapter 8). The identification of redundancy amongst variables has already been described (Section 3.5) and techniques have been discussed for the reduction of dimensionality as a step in data analysis. These procedures are unsupervised learning methods and thus do not use the property of most interest, the dependent variable. What about supervised learning methods? The obvious way in which supervised techniques may be used for the identification of important features is to examine the variables that are included in the models. Significant terms in MLR equations may point to important variables as may high loadings in PCR components or PLS latent variables. High loadings for variables in the latter are likely to be more reliable indicators of importance since the PLS variables are constructed to be highly correlated with the dependent variable.

LDA models may also be used to identify important variables but here it should be remembered that discriminant functions are not unique solutions. Thus, the use of LDA for variable selection may be misleading. Whatever form of supervised learning method is used for the identification of important variables it is essential to bear in mind one particular problem with supervised learning: chance effects. In order to reduce the probability of being misled by chance correlations it is wise to be conservative in the use of supervised learning techniques for variable selection.

7.5 Summary

This chapter has described some of the more commonly used supervised learning methods for the analysis of data; discriminant analysis and its relatives for classified dependent data, variants of regression analysis for continuous dependent data. Supervised methods have the advantage that they produce predictions, but they have the disadvantage that they can suffer from chance effects. Careful selection of variables and test/training sets, the use of more than one technique where possible, and the application of common sense will all help to ensure that the results obtained from supervised learning are useful.

References

Brill, J. H., Mayfield, H. T., Mar, T., and Bertsch, W. (1985). *Journal of Chromatography*, **349**, 31–8.

Clark, M. T., Coburn, R. A., Evans, R. T., and Genco, R. J. (1986). *Journal of Medicinal Chemistry*, **29**, 25–9.

Cramer, R. D., Patterson, D. E., and Bunce, J. D. (1988). *Journal of the American Chemical Society*, **110**, 5959–67.

Diana, G. D., Oglesby, R. C., Akullian, V., Carabateas, P. M., Cutcliffe, D., Mallamo, J. P., *et al.* (1987). *Journal of Medicinal Chemistry*, **30**, 383–8.

Dillon, W. R. and Goldstein, M. (1984). *Multivariate analysis methods and applications*, pp. 360–93. Wiley, New York.

Dunn, W. J., Wold, S., Edlund, U., Hellberg, S., and Gasteiger, J. (1984). *Quantitative Structure–Activity Relationships*, **3**, 134–7.

Dunn, W. J., Wold, S., and Martin, Y. C. (1978). *Journal of Medicinal Chemistry*, **21**, 922–30.

Goodford, P. J. (1985). *Journal of Medicinal Chemistry*, **28**, 849–57.

Hellberg, S., Wold, S., Dunn, W. J., Gasteiger, J., and Hutchings, M. G. (1985). *Quantitative Structure–Activity Relationships*, **4**, 1–11.

Kier, L. B. (1980). *Journal of Pharmaceutical Science*, **69**, 416–419.

Kubinyi, H. (ed.) (1993). *3D QSAR in drug design: theory, methods, and applications*. ESCOM, Leiden.

Lanteri, S., Conti, P., Berbellini, A., Centione, G., Polzonetti, A., Marassi, R., *et al.* (1988). *Journal of Chemometrics*, **3**, 293–9.

Li, R., Hansch, C., Matthews, D., Blaney, J. M., Langridge, R., Delcamp, T. J., Susten, S. S., *et al.* (1982). *Quantitative Structure–Activity Relationships*, **1**, 1–7.

Livingstone, D. J., Evans, D. A., and Saunders, M. R. (1992). *Journal of the Chemical Society—Perkin Transactions II*, 1545–50.

Malpass, J. A., Salt, D. W., Ford, M. G., Wynn, E. W. and Livingstone D. J. (1994) In *Advanced computer-assisted techniques in drug discovery*, (ed. H. Van de Waterbeemd), Vol 3 of *Methods and Principles in Medicinal Chemistry* (ed. R. Mannhold, P. Krogsgaard-Larsen and H. Timmerman) pp. 163-189. VCH, Weinheim

Martin, Y. C., Holland, J. B., Jarboe, C. H., and Plotnikoff N. (1974). *Journal of Medicinal Chemistry*, **17**, 409–13

Page, S. W. (1986). *Food Technology*, **Nov.**, 104–9.

Rose, V. S., Wood, J., and McFie, H. J. H. (1991). *Quantitative Structure–Activity Relationships*, **10**, 359–68.

Scarminio, I. S., Bruns, R. E., and Zagatto, E. A. G. (1982). *Energia Nuclear e Agricultura*, **4**, 99–111.

Stone, M. and Brooks, R. J. (1990). *Journal of the Royal Statistical Society Series B—Methodological*, **52**, 237–69.

Stouch, T. R. and Jurs, P. C. (1985a). *Journal of Chemical Information and Computer Sciences*, **25**, 45–50.

Stouch, T. R. and Jurs, P. C. (1985b). *Journal of Chemical Information and Computer Sciences*, **25**, 92–8.

Stuper, A. J. and Jurs, P. C. (1976). *Journal of Chemical Information and Computer Sciences*, **16**, 238–41.

Whalen-Pedersen, E. K. and Jurs, P. C. (1979). *Journal of Chemical Information and Computer Sciences*, **19**, 264–6.

Wilson, L. Y. and Famini, G. R. (1991). *Journal of Medicinal Chemistry*, **34**, 1668–74.

Wold, S. and Sjostrom, M. (1977). In *Chemometrics—theory and application*, (ed. B. R. Kowalski), p. 243. American Chemical Society, Washington D.C.

Wold, S. (1978). *Technometrics*, **20**, 397–405.

Zalewski, R. I. (1992). *Journal de Chimie Physique et de Physico-Chemie Biologique*, **89**, 1507–16.

8
Multivariate dependent data

8.1 Introduction

The last four chapters of this book have all been concerned with methods that handle multiple independent (descriptor) variables. This has included techniques for displaying multivariate data in lower dimensional space, determining relationships between points in N dimensions and fitting models between multiple descriptors and a single response variable, continuous or discrete. Hopefully, these examples have shown the power of multivariate techniques in data analysis and have demonstrated that the information contained in a data set will often be revealed only by consideration of all of the data at once. What is true for the analysis of multiple descriptor variables is also true for the analysis of multiple

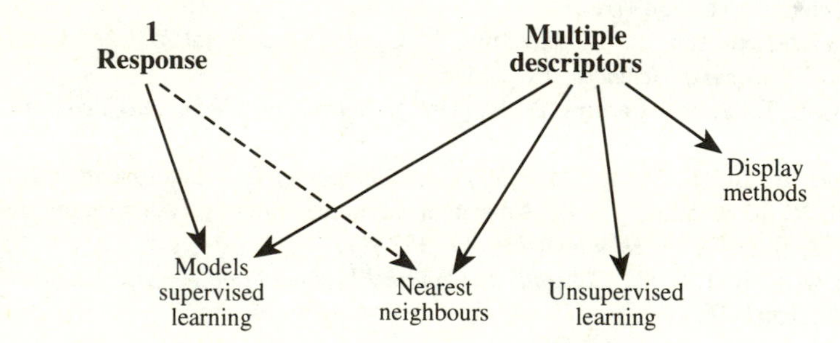

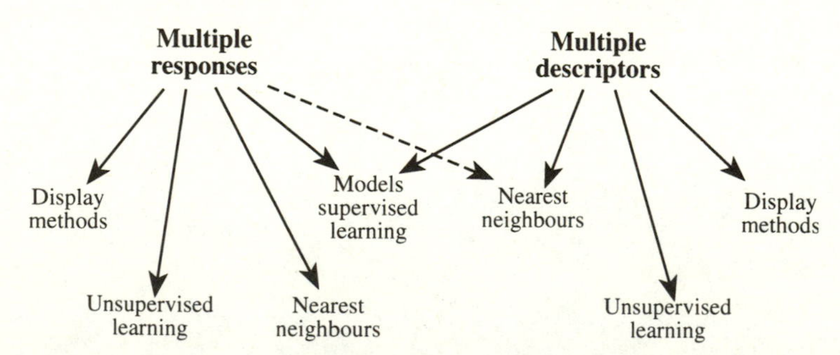

Fig. 8.1. Analytical methods that may be applied to one response with multiple descriptors and multiple responses with multiple descriptors.

response data. All of the techniques so far described for independent variables may also be applied to multiple dependent variables, as illustrated in Fig. 8.1. The following sections of this chapter demonstrate the use of principal components and factor analysis (FA) in the treatment of multiple response data, cluster analysis, and a method called spectral map analysis. The last section discusses the construction of models between multiple dependent and independent variable sets. It is perhaps worth pointing out here that the analysis of multivariate response data is even more unusual in the chemical literature than the multivariate analysis of independent data sets. This is probably not only a reflection of the unfamiliarity of many multivariate techniques but also of the way in which experiments are conducted. It is not uncommon to design experiments to have only one outcome variable, an easily determined quantity such as colour, taste, percentage inhibition, and so on. This demonstrates a natural human tendency to try to make a complicated problem (the world) less complicated. In many cases, our experiments do generate multiple responses but these are often discarded, or processed so as to produce a single 'number', because of ignorance of methods which can be used to handle such data. Perhaps the following examples will show how such data sets may be usefully treated.

8.2 Principal components and factor analysis

Compounds which are biologically active often show different effects in different tests, e.g., related receptor assays, as shown later, or in different species. A very obvious example of this is the activity of antibacterial compounds towards different types of bacteria. If a biological results table is drawn up in which each row represents a compound and each column the effect of that compound on a different bacterial strain, principal component analysis (PCA) may be used to examine the data. An example of part of such a data set, for a set of antimalarial sulphones and sulphonamides, is shown in Table 8.1. The application of PCA to this data gave two principal components which explained 88 per cent and 8 per cent of the variance respectively. Relationships between variables may be seen by construction of a loadings plot, as shown in Fig. 8.2, here the results of *M. lufu* and *E. coli* are seen to correspond to one another; the *plasmodium* data is quite separate. Pictures such as this can be useful in predicting the likely specificity of compounds and can also give information (potentially) on their mechanism of action. When different test results are grouped very closely together it might be reasonable to suppose that inhibition of the same enzyme is involved, for example. A multivariate response set need not be restricted to data from the same type of biological test. Nendza and Seydel (1988) have reported results of the toxic effects of a set of phenols

Table 8.1 Observed biological activity of 2',4'-substituted 4-aminodiphenylsulfones determined in cell-free systems (columns 1–3, I_{50} [μmol/L]) and whole cell systems (columns 4–6, I_{25}, MIC [μmol/L]) of plasmodia and bacterial strains as indicated (from Wiese *et al.* 1987, with permission of VCH, Weinheim)

No.	Compound	*P. berghei* I_{50}	*M. lufu* I_{50}	*E. coli* I_{50}	*E. coli* I_{25}	*M. lufu* MIC	*E. coli* MIC
1	4'-NH_2(DDS)	12.41	1.20	34.36	7.857	0.17	16.00
2	4'-OCH_3	–*	7.55	128.16	37.277	30.38	109.00
3	4'-NO_2	–	31.03	212.74	39.546	10.78	5.60
4	4'-H	104.00	12.06	116.53	33.456	51.44	45.00
5	4'-OH	32.23	1.50	34.92	5.867	20.06	22.50
6	4'-Cl	–	12.99	–	–	59.76	45.00
7	4'-$NHCOCH_3$	33.26	7.01	75.65	21.406	10.33	45.00
8	4'-Br	–	9.69	–	–	–	33.75
9	4'-$NHCH_3$	26.70	1.29	46.21	10.740	1.90	90.00
10	4'-NHC_2H_5	–	2.75	41.71	16.140	0.90	64.00
11	4'-CH_3	48.00	8.08	89.84	47.221	16.17	–
12	4'-$N(CH_3)_2$	21.51	1.76	–	42.374	12.66	–
13	4'-$COOCH_3$	147.00	11.75	149.29	123.950	13.73	–
14	4'-COOH	–	3.60	74.24	867.200	–	–
15	4'-$CONHNH_2$	76.510	12.72	155.39	36.158	27.46	64.00

* The blank entries indicate a compound not tested or missing a parameter.

and aniline derivatives on three bacterial, four yeast, two algae, one protoplast, and one daphnia system (Table 8.2). Loadings of these test systems on the first two principal components are shown in Table 8.3 and Fig. 8.3. The second principal component, which only explains nine per cent of the variance in the set of 26 compounds, appears to be mostly made up from the inhibition of algae fluorescence data. All of the test systems have a positive loading with the first component but inspection of the loadings plot in Fig. 8.3 shows that these variables fall into two groups

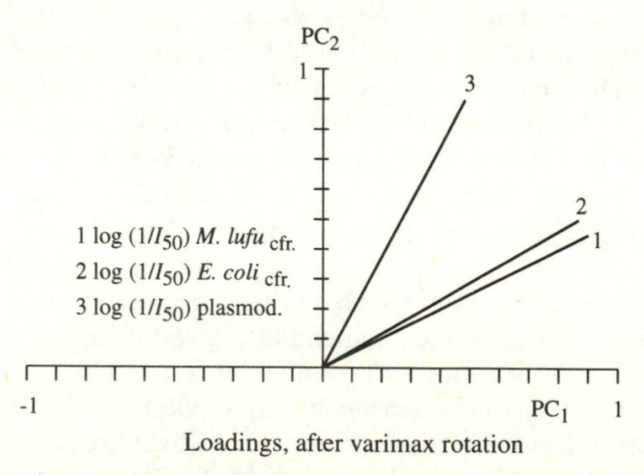

Fig. 8.2. Loadings plot for three biological responses on the first two principal component axes (from Wiese *et al.* (1987), with permission of VCH, Weinheim).

Table 8.2 Toxicity parameters of phenol and aniline derivatives (from Nendza and Seydel 1988, with permission of VCH, Weinheim)

		MIC *E. coli*	MIC M.169	I_{50} *E. coli*	I_{50} *Sacch. cer.*	I_{50} Purin uptake	I_{50} ATPase	I_{50} DEF	I_{50} Rubisco	I_{50} Algae	I_{50} Fluor	ED_{100} Daphnia
1	Phenol	11	5.7	7.0	14	26	11	45	6.3	6.4	0.33	–
2	4-Cl-phenol	2.0	0.72	1.02	1.8	4.6	4.4	8.6	1.2	0.091	0.055	0.086
3	2,3-Cl_2-phenol	0.36	0.26	0.28	0.40	0.84	0.71	2.1	–	0.086	0.14	0.047
4	2,4-Cl_2-phenol	0.36	0.36	0.33	0.38	1.1	0.81	1.7	0.33	0.10	0.005	0.031
5	2,6-Cl_2-phenol	1.4	0.72	0.72	0.70	2.0	1.6	3.0	5.0	0.67	0.11	0.37
6	2,4,5-Cl_3-phenol	0.045	0.045	0.042	0.033	0.15	0.13	0.55	–	0.017	0.0038	0.013
7	2,4,6-Cl_3-phenol	0.36	0.13	0.19	0.060	0.15	0.44	0.52	0.93	0.011	0.36	0.056
8	2,3,4,5-Cl_4-phenol	0.016	0.011	0.045	0.010	0.014	0.023	0.12	–	–	–	–
9	Cl_5-phenol	0.13	0.0056	0.12	0.087	0.0035	0.014	0.11	0.009	0.0023	0.001	0.0075
10	2-Br-phenol	1.4	1.4	0.75	1.9	4.3	2.3	8.4	1.9	0.28	0.14	0.10
11	Aniline	23	11	26.2	23	72	50	105	36	0.67	0.41	4.1
12	2-Cl-aniline	5.7	4.1	2.2	5.4	13	10	16	4.3	0.45	0.42	0.28
13	3-Cl-aniline	4.1	2.0	2.7	3.6	13	8.6	16	1.6	0.25	0.18	0.39
14	4-Cl-aniline	2.9	0.51	3.0	2.8	17	11	18	2.2	0.017	0.009	0.78
15	2,4-Cl_2-aniline	2.0	0.72	0.38	1.1	2.3	1.6	3.0	2.6	0.11	0.031	0.062
16	2,6-Cl_2-aniline	5.7	8.2	0.62	1.2	3.4	1.7	4.0	21	0.48	0.63	0.12
17	2-Br-aniline	2.9	4.1	1.1	3.6	8.7	7.8	11	5.2	0.38	0.15	0.17
18	2,4-$(NO_2)_2$-aniline	1.0	0.51	0.41	1.5	0.11	0.80	–	0.02	0.0049	0.093	0.11
19	4-CH_3-aniline	8.2	4.1	12.0	11	23	22	50	31	0.13	0.079	4.7
20	2,4-$(CH_3)_2$-aniline	8.2	0.72	2.9	6.4	9.0	11	18	7.9	0.096	0.077	1.7

Table 8.3. Loadings of the biological test systems on the first two principal components (from Nendza and Seydel 1988, with permission of VCH, Weinheim)

	Test system	PC_1	PC_2
1	MIC *E. coli*	0.89	0.14
2	MIC M. 169	0.91	0.21
3	I_{50} *E. coli*	0.89	−0.13
4	I_{50} *Sacch. cer.*	0.92	−0.24
5	I_{50} Purin	0.92	−0.16
6	I_{50} ATPase	0.95	−0.20
7	I_{50} DEF	0.90	−0.32
8	I_{50} Rubisco	0.88	0.24
9	I_{50} Algae	0.75	−0.27
10	I_{50} Fluorescence	0.65	0.70
11	I_{100} Daphnia (24 h)	0.89	−0.26

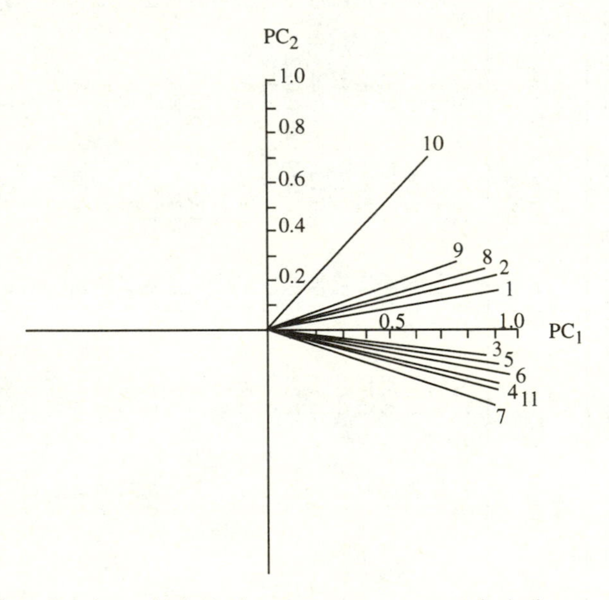

Fig. 8.3. Loadings for eleven biological test systems on two principal components; the identity of the test systems is given in Table 8.3 (from Nendza and Seydel 1988, with permission of VCH, Weinheim).

according to the sign of their loadings with the second component. Scores may be calculated for each of the principal components for each of the compounds and these scores used as a composite measure of biological effect. The scores for PC_1 were found to correlate well with log k′, a measure of hydrophobicity determined by HPLC, as shown in Fig. 8.4. Thus, it may be expected that lipophilicity is correlated with the individual results of each test system.

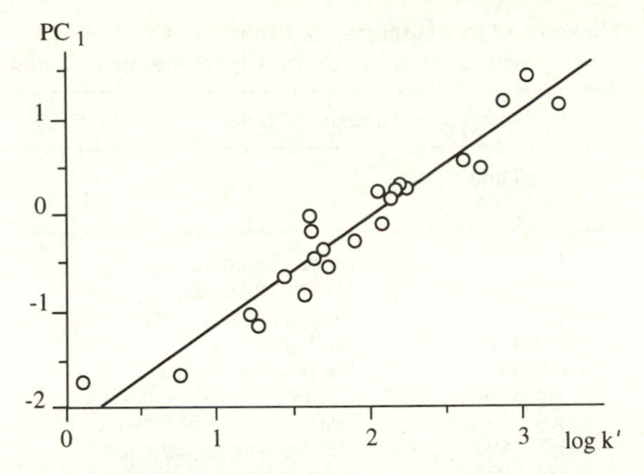

Fig. 8.4. Plot of PC_1 scores versus log k′ (from Nendza and Seydel 1988, with permission of VCH, Weinheim).

Response data from both *in vitro* and *in vivo* test systems may be combined and analysed by PCA or FA. This is a particularly useful procedure since *in vitro* tests are often expected (or assumed!) to be good models for *in vivo* results. A straightforward test of the simple correlation between any two experimental systems can of course be easily obtained but this may not reveal complex relationships existing in a response set. Figure 8.5 shows the results of a factor analysis of a combined set of *in vitro*, *in vivo*, and descriptor data, there being no reason why data sets should not be formed from a combination of dependent and independent data. The

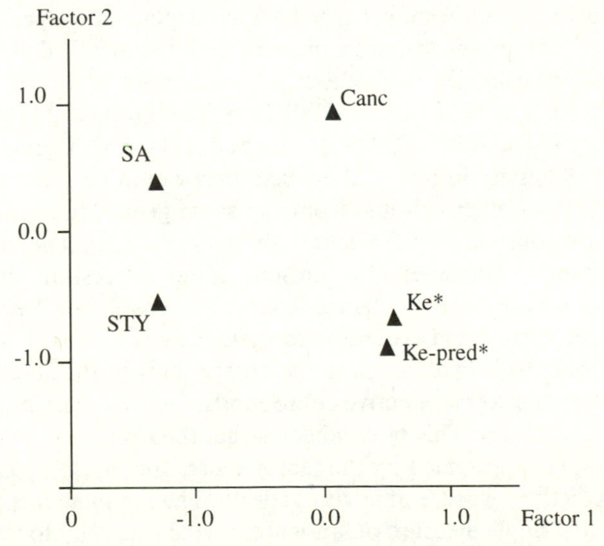

Fig. 8.5. Loadings plot from a factor analysis of *in vivo* and *in vitro* tests, and measured and calculated physicochemical descriptors (from Benigni *et al.* 1993 by permission of Oxford University Press).

Table 8.4. Cell lines used in the testing of antitumour platinum complexes (from Kuramochi *et al.* 1990, with permission of the Pharmaceutical Society of Japan)

	Tumour cell lines	
Number on Fig. 8.6	Cell line	Origin
1	L1210	Mouse leukemia
2	P388	Mouse leukemia
3	LL	Mouse lung carcinoma
4	AH66	Rat hepatoma
5	AH66F	Rat hepatoma
6	HeLa S_3	Human cervical carcinoma
7	KB	Human nasopharyngeal carcinoma
8	HT-1197	Human bladder carcinoma
9	HT-1376	Human bladder carcinoma

factor plot shows that a calculated (Ke*-pred) and experimental (Ke*) measure of chemical reactivity fall close together, while in another part of factor space, experimental Ames test (STY) results (*in vitro*) and a predicted measure of mutagenicity (SA) are associated. Both of these associated sets of responses are separated from the *in vivo* measures of rat carcinogenicity (Canc), which they are expected to predict (at least to some extent). Another example of the use of factor analysis in the treatment of multiple response data involved the antitumour activity of platinum complexes against the tumour cell lines shown in Table 8.4 (Kuramochi *et al.* 1990). Three factors were extracted which explained 84 per cent of the variance of 52 complexes tested in these nine different cell lines. A plot of the rotated* factor loadings for the first two factors is shown in Fig. 8.6 where it can be seen that the tests fall into four groups: AH66F, L1210, AH66; HeLa, P388, KB; HT-1197, LL; and HT-1376. Compounds that exhibit a given activity in one of these cell lines would be expected to show similar effects in another cell line from the same group, thus cutting down the need to test compounds in so many different cell lines. The factor scores for the platinum complexes also present some interesting information. Figure 8.7 shows a plot of the factor 2 scores versus factor 1 scores where the points have been coded according to their activity against L1210 *in vivo*. Factor 2 appears to broadly classify the compounds in that high scores on factor 2 correspond to more active compounds; factor 1 does not appear to separate the complexes. This plot indicates that the results obtained in the *in vitro* cell lines, as represented by the factor scores for the complexes, can be used as a predictive measure of *in vivo* activity. The factor scores can also be used as a simple single measure of 'anticarcinogenic' activity for use in other

* Simplified by varimax rotation, see Section 4.2.

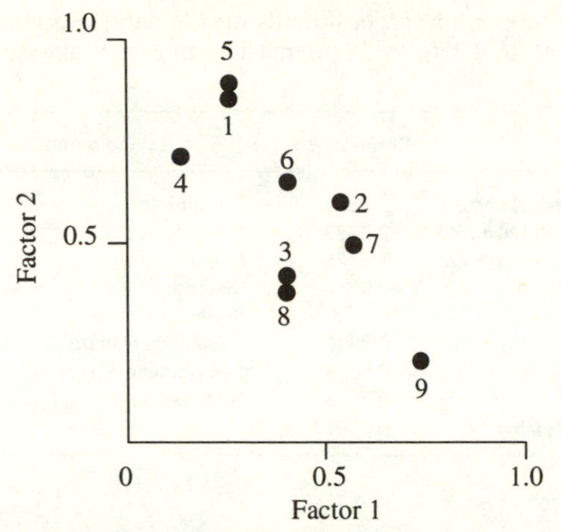

Fig. 8.6. Loadings plot from a factor analysis of the activity of platinum complexes in a set of nine different tumour cell lines. Cell lines are identified by the numbers in Table 8.4 (from Kuramochi *et al.* 1990, with permission of the Pharmaceutical Society of Japan).

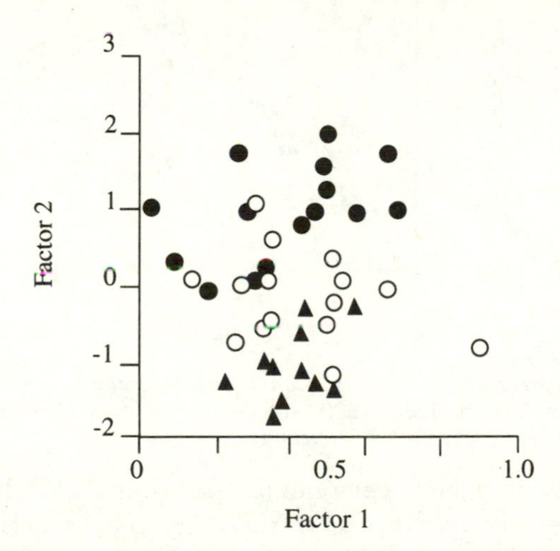

Fig. 8.7. Scores plot of the first two factor axes for high activity (●), moderate activity (○), and low activity (▲) complexes against L1210 (from Kuramochi *et al.* 1990, with permission of the Pharmaceutical Society of Japan).

methods of analysis. The complexes were made up of carrier ligands and leaving groups and the activity contribution of each type of ligand or leaving group to the factor 2 scores was evaluated by the Free–Wilson method (Section 6.3.3). The results of this analysis are shown in Table 8.5 where the higher positive values indicate greater contribution to the factor scores, and hence *in vivo* activity. Factor scores estimated by the Free–Wilson method are well correlated to the measured factor scores as shown in Fig. 8.8. Factor scores or principal component scores can thus be used as new variables representing a set of multiple responses, just as they have

Table 8.5. Contributions of carrier ligands and leaving groups to factor 2 scores (from Kuramochi *et al.* 1990, with permission of the Pharmaceutical Society of Japan)

Carrier ligand	Contribution	Leaving group	Contribution
1-(Aminomethyl)cyclohexylamine	2.5694	Tetrachloro	0.0599
1-(Aminomethyl)cyclopentylamine	2.2497	Dichloro	0.0
1,1-Diethylethylenediamine	1.7729	Oxalato	−0.1479
1,2-Cyclohexanediamine	1.6738	Malonato	−0.2581
1,4-Butanediamine	0.7494	Sulfato	−0.2438
1,1-Dimethylethylenediamine	0.7370	2-Methylmalonato	−0.3393
3,4-Diaminotetrahydropyran	0.5084	Cyclobutane-1,1-dicarboxylato	−0.8787
Diamine	0.0	Dihyroxydichloro	−1.4111
N,N-dimethylethylenediamine	−0.08957		

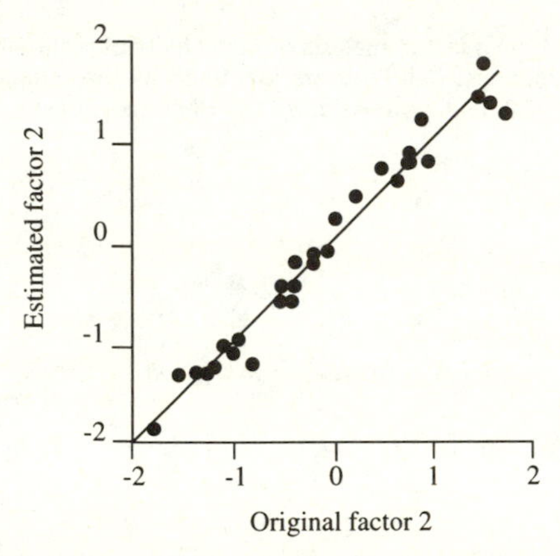

Fig. 8.8. Plot of factor 2 scores estimated from a Free–Wilson analysis versus experimental factor 2 scores (from Kuramochi *et al.* 1990, with permission of the Pharmaceutical Society of Japan).

been used as new variables representing a set of multiple descriptors. Section 8.5 describes the construction of regression-like models using multiple response and descriptor data.

8.3 Cluster analysis

We have already seen, in Section 5.4 on cluster analysis, the application of this method to a set of multiple response data (Fig. 5.9). In this example the biological data consisted of 12 different *in vivo* assays in rats so the y (dependent) variable was a 40 (compounds) by 12 matrix. These compounds exert their pharmacological effects by binding to one or more of the neurotransmitter binding sites in the brain. Such binding may be characterized by *in vitro* binding experiments carried out on isolated tissues and a recent report (Testa *et al.* 1989) lists data for the binding of 21 neuroleptic compounds to the following receptors:

α-noradrenergic, α1 and α2
β-noradrenergic, β(1 + 2)
dopaminergic, D1 and D2
serotoninergic, 5HT1 and 5HT2
muscarinic, M(1 + 2)
histaminic, H1
opioid
Ca^{2+} channel
serotonin uptake

A dendrogram showing the similarities between these compounds is given in Fig. 8.9 where it can be seen that there are three main clusters. Cluster A, which is quite separate from the other clusters, contains the benzamide

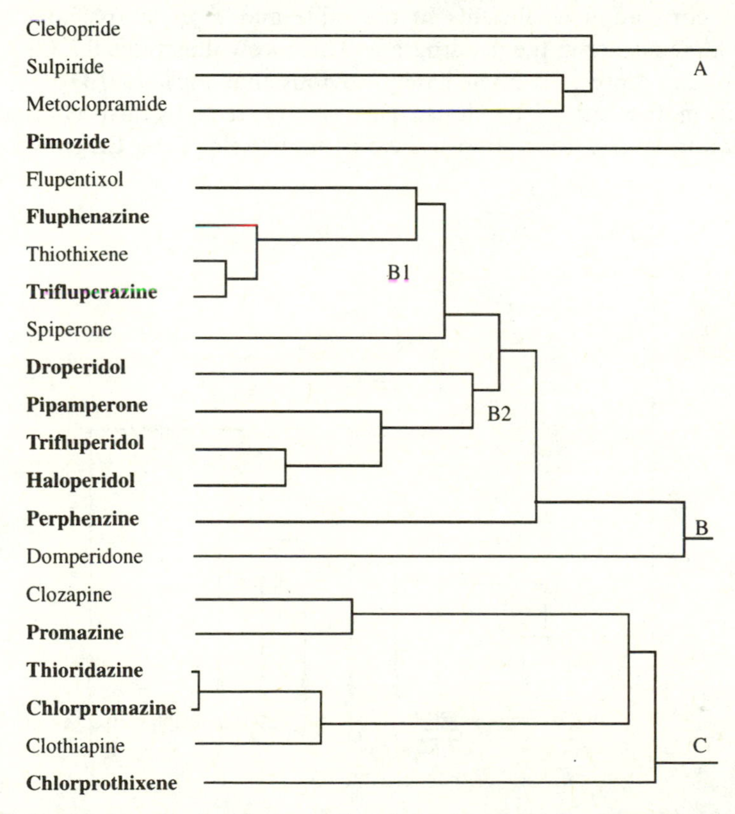

Fig. 8.9. Dendrogram showing the associations between neuroleptic compounds binding to 14 receptors (from Testa *et al.* 1989, with permission of Plenum Publishing Corporation).

drugs. The compounds in cluster B, made up of two subgroups of different chemical classes, are characterized by higher affinity for D2 and 5HT2 receptors compared to D1, α1 and 5HT1 and have no activity at muscarinic receptors. Cluster C contains compounds with high α1 affinity, similar D1 and D2 affinity and a measurable affinity for muscarinic receptors (all other compounds have $-\log IC_{50} < 4$ for this receptor). Compounds shown in bold in Fig. 8.9 were also present in the data set shown in Fig. 5.9 and in some cases fall into similar clusters, e.g., fluphenazine and trifluperazine in cluster B1 are present in the same cluster in Fig. 5.9. This shows the anticipated result that *in vitro* receptor binding data may be used to explain *in vivo* pharmacological results.

Cluster analysis can also be used to show relationships between variables, dependent or independent. Again, as was said in Section 5.4, a data set of *p* objects in an *N*-dimensional space can also be viewed as a set of *N* objects in a *p*-dimensional sample space. This is demonstrated in Fig. 8.10 for the biological test results shown in Table 8.6 (Mazerska *et al.* 1990). All of the test results are related to one another to some extent as shown by the correlation coefficients in the table and more graphically by the single large cluster in the dendrogram. This nicely illustrates the utility of a dendrogram since it is immediately obvious that $\log TD_{50}/ED_{50}$ is separated from the rest of the tests. This can be seen by inspection of the correlation matrix, for example, the bottom line ($\log TD_{50}/ED_{50}$) where the

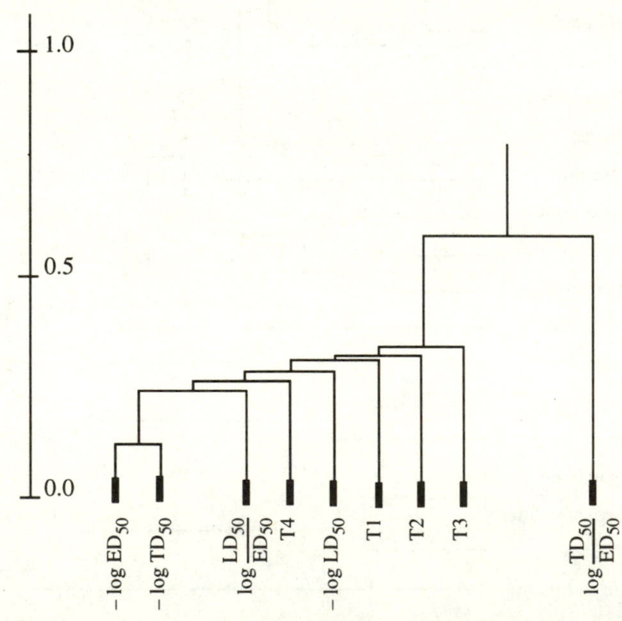

Fig. 8.10. Dendrogram of the associations between the biological tests shown in the correlation matrix in Table 8.6 (from Mazerska *et al.* 1990, with permission of Macmillan).

Table 8.8. Rotated factor loadings for ten biological response variables* (from Livingstone *et al.* 1988, with permission of Elsevier Science)

	Killing activity (KA)	Time onset	Time maximum frequency (LTMF)	Minimum threshold conc. (MTC)	Max. frequency of AP	Time to block	Slope	Intercept	Max. burst freq. (LBF)	Knock-down (KDA)
F1	−0.93	0.84	0.88	0.71	0.74	0.79	−0.16	0.21	0.06	−0.50
F2	0.28	−0.10	−0.29	−0.57	0.39	0.04	0.23	−0.18	0.90	0.72
F3	0.01	0.15	−0.10	0.00	−0.31	0.10	0.93	0.94	−0.04	0.16

* *In vivo* data (KA and KDA) measured using *Musca domestica* L., *in vitro* data measured using isolated haltere nerves.

highest value is 0.494, but this is not as immediately apparent as the dendrogram. The dendrogram also clearly shows the unfortunate similarity between the dose required for therapeutic effect (ED_{50}) and the dose which shows toxicity (TD_{50}).

8.4 Spectral map analysis

Spectral map analysis (SMA) was briefly mentioned in Chapter 4 with an example of a spectral map shown in Fig. 4.3. The reason for the development of this technique was the use of activity spectra to represent the activity of compounds in several different pharmacological (usually, but not necessarily) tests. An example of activity spectra for four α-agonists in six tests on rats is shown in Fig. 8.11. The three tests A, B, and C are *in vivo* antidiarrhoeal, diuretic, and antiptotic respectively; tests D, E, and F are *in vitro* binding (Lewi, 1989). The similarities between the two compounds in each pair can immediately be seen. The shape of their activity profiles, or spectra, demonstrate that guanabenz and nordephrine have similar selectivity for the six test results shown and simply differ in terms of their potency. Similarly, amidephrine and methoxamine have virtually identical spectra with amidephrine being the more active (the activity scale is log $1/C$ to cause a standard effect).

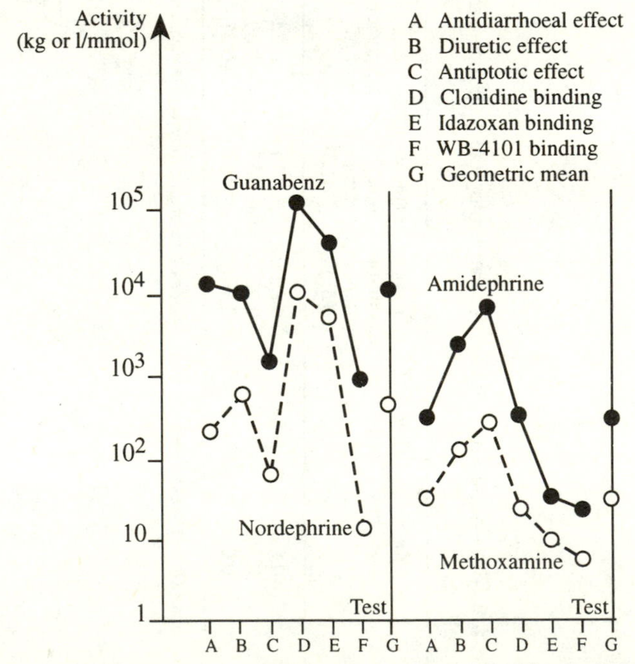

Fig. 8.11. Activity spectra for four α-agonists (from Lewi 1989, with permission of Elsevier Science).

The distinction between activity and specificity has been likened by Lewi (1989) to the difference between size and shape. As an example, he suggested the consideration of a table of measurements of widths and lengths of a collection of cats, tigers, and rhinoceroses. With respect to size, tigers compare well with rhinoceroses, but with respect to shape, tigers are classified with cats. In order to characterize shape it is necessary to compare width/length ratios with their geometric mean ratio. This is called a contrast and can be defined as the logarithm of an individual width/length ratio divided by the mean width/length ratio. From the table of animal measurements, a positive contrast would indicate a rhinoceros (width/length ratio greater than the mean width/length ratio), whereas a negative contrast could indicate a cat or a tiger. If the contrast is near zero, this may mean an overfed cat, a starving rhinoceros, or some error in the data.

For the compounds and tests represented by the activity spectra shown in Fig. 8.11, a single major contrast allowed the correct classification to be made visually. However, when more than one important contrast is present in the data it becomes difficult, if not impossible, to identify these contrasts by simple inspection of activity spectra. This is where SMA comes in as it is a graphical method for displaying all of the contrasts between the various log ratios in a data table. The SMA process consists of the following steps.

1. Logarithmic transformation of the data.
2. Row centring of the data (subtraction of the mean activity of a compound in all tests).
3. Column centring of the data (subtraction of the mean activity of all compounds in one test).
4. Application of factor analysis to the doubly centred data.
5. Application of scaling* to allow both factor scores and loadings to be plotted on the same plot (called a biplot).

The end result of this procedure is the production of a biplot in which the similarities and differences between compounds, tests and compound/tests can be seen.

The activity spectra shown in Fig. 8.11 come from a data set of 18 α-agonists tested in the six different tests. Application of SMA to this data set produced three factors which explained 48, 40, and 8 per cent respectively of the variance of contrasts. The factor scores and loadings were used to produce the biplot shown in Fig. 8.12, the x and y axes representing the first and second factors, respectively. At first sight it appears that this figure is horribly complicated but it does contain a great

---

* Various scaling options can be applied to produce biplots, see Lewi (1989) for details.

deal of information and the application of a few simple 'rules' does allow a relatively easy interpretation of the plot. The rules are as follows.

1. Circles represent compounds and squares represent tests. For example, oxymetazoline and clonidine binding in Fig. 8.12.
2. Areas of circles and squares are proportional to the mean activity of the compounds and tests. For example, the mean activity of compounds in the antidiarrhoeal and diuretic tests are similar.
3. The position of compounds and tests are defined by their scores and loadings on the first two factors. Where compounds are close together they have similar activity profiles; where tests are close together they give similar results for the same compounds; where compounds are close to tests they have high specificity for those tests. Compounds and tests that have little contrast lie close to the centre of the map. For example nordephrine and guanabenz are close together on Fig. 8.12 and have similar activity profiles as shown in Fig. 8.11. These compounds are close to the square symbol for clonidine binding (test D in Fig. 8.11).
4. The third most significant factor is coded in the thickness of the contour around a symbol—a thick contour indicates that the symbol is above the plane of the plot, a thin contour that it lies below it.
5. Compounds and tests that are not represented in the space spanned by these three factors are represented by symbols with a broken line contour (none in this example).

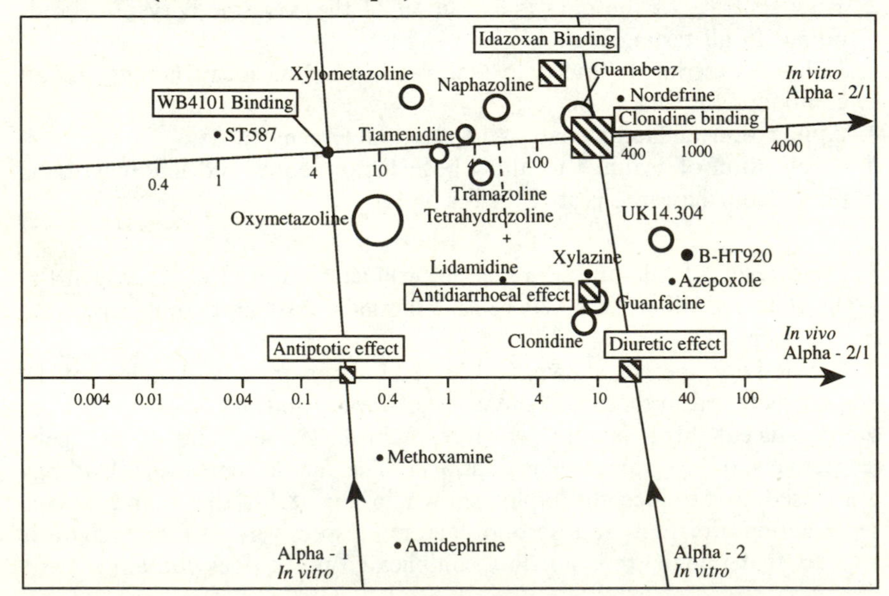

Fig. 8.12. Spectral map of factor scores and loadings (from Lewi 1989, with permission of Elsevier Science).

6. An axis of contrast can be defined through any two squares representing tests.

There are many useful features of a data set that can be revealed in a spectral map such as this and we can see that it is consistent with the patterns shown in the activity spectra (Fig. 8.11) since these two pairs of compounds are grouped together on the plot. The use of the thickness of symbol contours to denote the third factor is perhaps not very successful but modern computer graphics would allow easy display of such a plot in three dimensions. In this particular example, the third factor is probably not very important since it only describes eight per cent of the variance of contrasts. SMA is clearly a useful method for the analysis of any chemical problem in which a set of compounds is subjected to a battery of tests which produce some quantitative measure of response. It offers the advantage of a simultaneous display of the relationships between both tests and compounds.

8.5 Models for multivariate dependent and independent data

Chapters 6 and 7 described the construction of regression models (MLR, PCR, PLS, and continuum regression) in which a single dependent variable was related to linear combinations of independent variables. Can these procedures be modified to include multiple dependent variables? One fairly obvious way to take account of at least some of the information in a multivariate dependent set is to carry out PCA or FA on the data and use the resulting scores to construct regression models.

The biological activity data for the nitro-9-aminoacridines shown in Table 8.6 and Fig. 8.10 were analysed by PCA to give two principal components explaining 81 per cent and 8 per cent of the variance in the set (Mazerska *et al.* 1990). The loadings of the biological tests on these two principal components are shown in Table 8.7. All of the tests appear to have a high positive loading on the first component and this PC was interpreted as being a measure of 'general biological activity'. The main variable loading onto PC_2 was the therapeutic index, although this loading is quite small (0.531). Thus the second PC might be interpreted as a measure of 'selectivity'. The compounds were described by log P and a number of topological descriptors, and regression equations were sought between the principal component scores and these physicochemical parameters. The first component was well described by a parabolic relationship with modified (see paper for details) log P values

Table 8.7. Loadings of the two significant principal components after varimax rotation (from Mazerska *et al.* 1990, with permission of Macmillan)

Biological tests*	Loadings for	
	PC_1	PC_2
T1	0.903	−0.124
T2	0.863	−0.252
T3	0.812	−0.386
T4	0.934	0.114
p LD_{50}	0.918	−0.252
p ED_{50}	0.971	0.195
p TD_{50}	0.967	0.151
log LD_{50}/ED_{50}	0.834	0.531

* The symbols of the biological tests have the same meaning as in Table 8.6.

$$PC_1 = -0.258(\pm 0.03)\log P_*^2 - 0.80(\pm 0.08)\log P_* + 0.30(\pm 0.11) \quad (8.1)$$

$$n=28 \quad R=0.97 \quad s=0.23 \quad F=201$$

The values in brackets are the standard errors of the regression coefficients and it should be noted that R is quoted not R^2 as is more usual for multiple regression equations. The second PC was less well described by a shape parameter based on molecular connectivity (^2K) and an indicator variable.

$$PC_2 = 0.028(\pm 0.018)^2K - 0.122(\pm 0.07)I_{N2N} - 0.124(\pm 0.101) \quad (8.2)$$

$$n=28 \quad R=0.762 \quad s=0.07 \quad F=17.33$$

Equation (8.2) only describes 60 per cent of the variance in PC_2 and the high standard error for the shape descriptor term casts some doubt on the predictive ability of the equation. However, it is hoped that these two equations demonstrate the way in which regression models for multivariate dependent data can be generated by means of PCA.

An alternative method for the construction of regression type models for multivariate response sets is a technique known as **canonical correlation analysis (CCA)**. CCA operates by the construction of a linear combination of q responses

$$W_1 = a_{1,1}Y_1 + a_{1,2}Y_2 + \cdots a_{1,q}Y_q \quad (8.3)$$

and a linear combination of p descriptors

$$Z_1 = b_{1,1}X_1 + b_{1,2}X_2 + \cdots b_{1,p}X_p \quad (8.4)$$

where the coefficients in eqns (8.3) and (8.4) are chosen so that the pairwise correlation between W_1 and Z_1 is as large as possible. W_1 and Z_1 are referred to as canonical variates and the correlation between them as the canonical correlation. Further pairs of canonical variates may be calculated until there are as many pairs of variates as the smaller of p and q. Important features of the canonical variates are listed below.

1. They are generated in descending order of importance, i.e., canonical correlation coefficients decrease for successive pairs of variates.
2. Successive pairs of variates are orthogonal to one another.

There should be a ring of familiarity about this description of canonical correlation when we compare it with PCA or FA and, indeed, CCA can be considered as a sort of joint PCA of two data matrices. The canonical variates are orthogonal to one another, as are principal components and factors, and they are generated in decreasing order of importance, although for CCA importance is judged by the correlation between canonical variates, not the amount of variance they explain which is the criterion used for principal components and factors. In this respect, CCA can also be seen to be akin to PLS since the latent variables in PLS are constructed so as to explain variance and maximize their correlation with a dependent variable. One other similarity which should be pointed out, which may not be immediately obvious, is the relationship between CCA and multiple linear regression (MLR). MLR normally involves a single response variable, Y, and thus we can write

$$W_1 = Y \tag{8.5}$$

and this gives rise to one pair of canonical variates

$$Z_1 = b_{1,1}X_1 + b_{1,2}X_2 + \cdots b_{1,p}X_p \tag{8.6}$$

where the coefficients are chosen to maximize the correlation between Z_1 and $W_1\,(Y)$. Equation (8.6) shows us that MLR can be viewed as a special case of canonical correlation analysis. Before moving on to an example of CCA, it is perhaps worth pointing out some obscuring jargon which is sometimes used to 'explain' the results of application of the technique. The canonical variates produced by linear combinations of the response set (W_1, W_2, and so on) are often called the nth canonical variate of the first set. Similarly, the linear combinations of the descriptor set (Z_1, Z_2, and so on) can be called the nth canonical variate of the second set. Thus, descriptions like 'the second canonical variate of the first set', and 'the first canonical variate of the second set' can confuse when what is meant is W_2 and Z_1 respectively.

Application of CCA to the neurotoxic effects of pyrethroid analogues has been reported by Livingstone and co-workers (1988). Factor analysis

Table 8.6. Correlation matrix for biological test results for 41 nitro-9-aminoacridine derivatives (from Mazerska *et al.* 1990, with permission of Macmillan)

	T1	T2	T3	T4	p LD_{50}	p ED_{50}	p TD_{50}	log LD_{50}/ED_{50}	log TD_{50}/ED_{50}
T1	1.000								
T2	0.781	1.000							
T3	0.724	0.684	1.000						
T4	0.800	0.846	0.686	1.000					
p LD_{50}	0.849	0.811	0.797	0.788	1.000				
p ED_{50}	0.847	0.754	0.721	0.896	0.877	1.000			
p TD_{50}	0.816	0.777	0.738	0.886	0.877	0.980	1.000		
log LD_{50}/ED_{50}	0.686	0.570	0.523	0.822	0.608	0.914	0.881	1.000	
log TD_{50}/ED_{50}	0.433	0.153	0.154	0.374	0.289	0.447	0.260	0.494	1.000

T1—inhibition of *S. cerevisiae.*
T2—inhibition of germination of *L. sativum* sprouts.
T3—dehydrogenase activity inhibition in mouse tumour cells.
T4—inhibition of HeLa cell growth.
p LD_{50}—measure of acute toxicity in mice.
p ED_{50}—measure of antitumour activity in mice.

of a set of computed physicochemical properties for these compounds has been discussed in Section 5.3 of Chapter 5 (Table 5.9). Analysis of two *in vivo* responses, knockdown (KDA) and kill (KA), and eight *in vitro* responses by factor analysis led to the identification of three significant factors as shown in Table 8.8. Factor 1 is associated with both killing and knockdown activity (factor loadings of −0.93 and −0.5 respectively) and several neurotoxicological responses, while factor 2 is mostly associated with knockdown. The third factor, although judged to be statistically significant, is almost entirely composed of the slope and intercept of the *in vitro* dose response curves and was judged not to have neurotoxicological significance. From this factor analysis of the combined *in vitro* and *in vivo* data, it was possible to identify three *in vitro* responses which had high association with the *in vivo* data, in itself a useful achievement since it is usually easier (and more accurate) to acquire *in vitro* data. The responses were the logarithm of the time to maximum frequency of action potentials (LTMF) which has a high loading on factor 1, the logarithm of the maximum burst frequency (LBF) which has a high loading on factor 2 and the logarithm of the minimum threshold concentration (MTC) which loads on to both factors. These three responses were used to select a subset of six physicochemical properties (see paper for details) which describe the pyrethroid analogues. Canonical correlation analysis of this set of three *in vitro* responses and six physicochemical descriptors gave rise to two pairs of significant canonical variates with correlations of 0.91 and 0.88 respectively. The first canonical variate had a high association with MTC and thus might be expected to be a good predictor for both knockdown and kill. The second canonical variate had a significant association with LBF and thus should model knockdown alone. An advantage of CCA is the simultaneous use of both response and descriptor data but this can also be a disadvantage. In order to make predictions for a new compound it is necessary to calculate values for its physicochemical properties and then to combine these using the coefficients from the canonical correlation to produce descriptor scores (scores for the second set). These scores are equated by the canonical correlation to response scores (scores for the first set) but of course the scores are linear combinations of the individual responses. In order to predict an individual response for a new compound, it is necessary to obtain measured values for the other responses. This may be advantageous if one of the response scores is difficult or expensive to measure, for example, an *in vivo* response. The ability to predict *in vivo* responses from *in vitro* data may be a significant advantage in compound design. It may also be possible to make an estimate for an individual response by making assumptions about the values of the remaining response variables. The extra complexity of an approach such as CCA may be a disadvantage but it also offers a number of advantages.

8.6 Summary

This chapter has shown how multivariate dependent data, from multiple experiments or multiple results from one experiment, may be analysed by a variety of methods. The output from these analyses should be consistent with the results of the analysis of individual variables and in some circumstances may provide information that is not available from consideration of individual results. In this respect the multivariate treatment of dependent data offers the same advantages as the multivariate treatment of independent data. The simultaneous multivariate analysis of response and descriptor data may also be advantageous but does suffer from complexity in prediction.

References

Benigni, R., Cotta-Ramusino, M., Andreoli, C., and Giuliani, A. (1993). *Carcinogenesis*, **13**, 547–53.

Kuramochi, H., Motegi, A., Maruyama, S., Okamoto, K., Takahashi, K., Kogawa, O., *et al.* (1990). *Chemical and Pharmaceutical Bulletin*, **38**, 123–7.

Lewi, P. J. (1989). *Chemometrics and Intelligent Laboratory Systems*, **5**, 105–16.

Livingstone, D. J., Ford, M. G., and Buckley, D. S. (1988). In *Neurotox 88: Molecular basis of drug and pesticide action* (ed. G. G. Lunt), pp. 483–95. Elsevier, Amsterdam.

Mazerska, Z., Mazerska, J., and Ledochowski, A. (1990). *Anti-cancer Drug Design*, **5**, 169–87.

Nendza, M. and Seydel, J. K. (1988). *Quantitative Structure–Activity Relationships*, **7**, 165–74.

Testa, R., Abbiati, G., Ceserani, R., Restelli, G., Vanasia, A., Barone, D., *et al.* (1989). *Pharmaceutical Research*, **6**, 571–7.

Wiese, M., Seydel, J. K., Pieper, H., Kruger, G., Noll, K. R., and Keck, J. (1987). *Quantitative Structure–Activity Relationships*, **6**, 164–72.

9
Artificial intelligence

9.1 Introduction

The workings of the human mind have long held a fascination for man, and we have constantly sought to explain how processes such as memory and reasoning operate, how the senses are connected to the brain, what different physical parts of the brain do, and so on. Almost inevitable consequences of this interest in our own minds are attempts to construct devices which will imitate some or all of the functions of the brain, if not artificial 'life' then at least artificial intelligence. It might be thought that we have already achieved this goal when some of the awesome computing tasks, such as weather forecasting, that are now carried out quite routinely (and surprisingly accurately) are considered. Nowadays, for example, even a simple electrical appliance like the humble toaster is likely to contain a microchip 'brain'. Computers, of course, have revolutionized artificial intelligence (AI) research, so much so that devices are now being built which are models, albeit limited, of the physical organization and 'wiring' of the brain. These systems are known as artificial neural networks (ANN) and they have proved to be so remarkably successful that they have found application in a very diverse set of fields as shown in Table 9.1. The use of ANN in the analysis of chemical data is discussed in Section 9.3.

AI research has already provided the concepts of supervised and unsupervised learning to data analysis, and these have proved useful in the classification of analytical methods and to alert us to the potential danger of chance effects. But what of the application of AI techniques themselves

Table 9.1. Applications of artificial neural networks

'Reasoning'*	Process control
Verification of handwriting on cheques	Traffic control on underground stations
Identification of faces for a security system	Control of a magnetic torus for nuclear fusion
Credit ratings	Control of a chemical plant
Stock market forecasting	Control of a nuclear reactor
Drug detection at airports	
Grading pork for fat content	

* Including pattern recognition.

to the analysis of chemical data? The linear learning machine, or discriminant analysis, (see Section 7.2.1) is an AI method that is used in data analysis, but perhaps the most widely used AI technique is a method called expert systems. There are various flavours of expert systems and some authors apply the term to models, e.g., regression models, which have been derived from a particular set of data. In this chapter, the term expert system is used only to mean a procedure which has been created by some human expert or panel of experts. The expert systems described here do not necessarily have a role in data analysis but they do have a role in chemistry, for example, to calculate octanol/water partition coefficients and predict toxicity. The next section of this chapter deals with expert systems and the final section contains miscellaneous examples of the use of AI methods in chemistry.

9.2 Expert systems

The heart of any expert system consists of a set of rules, sometimes referred to as a rule base or knowledge base, which has been put together by 'experts'. The question of course arises of how to define (or find) the experts, but for the purposes of this discussion an expert is any human who has an opinion on the particular problem to be solved! Expert systems are usually, but not necessarily, implemented on a computer (see, for example, the structural alert system later in this section).

A simple example from the FOSSIL (Frame Orientated System for Spectroscopic Inductive Learning) system, which aims to identify chemicals from spectral data, may illustrate the expert system approach (Ayscough *et al.* 1987). This system contains, in its knowledge base, information about NMR spectra, infrared spectra, mass spectra, and ultraviolet spectra along with spectral heuristics (most suitable technique for assignment of particular structural properties, etc.) and molecular structures, functional groups, etc. One of the rules in the infrared section concerns the assignment of a methyl stretch—are there peaks in the IR spectrum in the region 2950–2975 cm^{-1} and 2860–2885 cm^{-1}? This information may be obtained from the user of the system by a prompt (the program asks a question) or might come from the automatic interpretation of a spectrum. A positive answer to this question would indicate the presence of a methyl group, and it is easy to see how similar rules in this and other parts of the knowledge base would allow assignment of various structural features. Some spectral information, of course, may indicate the presence of several alternative molecular features and here it is necessary to construct logical queries using operators such as IF, AND, NOT, THEN, ELSE, etc. The information required to satisfy such a query may be contained entirely in one spectrum or may need to come from other sources such as NMR and UV. Once the molecular features have been recognized, the construction of a molecule from them is not a trivial task since there will be many ways in which such fragments can be put together. A commonly adopted approach to this type of problem in expert systems is the construction of a decision tree which contains a number of connected nodes. Each node represents a logical question which has two or more answers and which dictates how the expert system proceeds with a query, in this example, the construction of a molecule from the identified features. Further discussion of expert systems can be found in the book by Cartwright (1993) and the review by Jakus (1992).

9.2.1 Log *P* calculation

One of the earliest applications of expert systems in the field of QSAR was the development of calculation schemes for octanol/water partition coefficients. Although the early work with π constants had shown that they were more or less additive, a number of anomalies had been identified. In addition, in order to calculate log *P* values from π constants it is necessary to have a measured log *P* for the parent and this, of course, is often unavailable. One approach to the question of how to calculate log *P* from chemical structure is to analyse a large number of measured log *P* values so as to determine the average contribution of particular chemical fragments (Nys and Rekker 1973). The fragment contributions constitute the rules of the expert system, extra rules being supplied in the form of correction factors. Operation of this expert system consists of the following few simple steps.

1. Break down the chemical structure into fragments that are present in the fragment table.
2. Identify any correction factors that are needed.
3. Add together the fragment values and apply the necessary correction factors to obtain a calculated log *P*.

Table 9.2 gives an example of some of the fragment contributions for this method; interestingly the correction factors always appeared to adopt the same value (0.28) or multiples of it. This was originally given the perhaps unfortunate name 'the magic constant'. The Rekker system of log *P* calculation was based on a statistical analysis of a large number of measured partition coefficients and can thus be called a reductionist approach. An alternative procedure was proposed by Hansch and Leo (1979) which involved a small number of 'fundamental' fragment values derived from very accurate partition coefficient measurements of a relatively small number of compounds. This technique, which can be viewed as a constructionist approach, requires a larger number of correction factors as

Table 9.2. Some fragment constants and correction factors for the Rekker log *P* prediction method[a] (from Rekker and de Kort 1979, with permission of the *European Journal of Medicinal Chemistry*)

Fragment[b]	Value	Fragment	Value
Br (Al)	0.249	COOH (Al)	−0.938
Br (Ar)	1.116	COOH (Ar)	−0.071
Cl (Al)	0.057	$CONH_2$ (Al)	−1.975
Cl (Ar)	0.924	$CONH_2$ (Ar)	−1.108
NO_2 (Al)	−0.920	CH_3	0.701
NO_2 (Ar)	−0.053	C_6H_5	1.840
OH (Al)	−1.470	pyridinyl	0.520
OH (Ar)	−0.314	indolyl	1.884

Correction factors

Type	Value
Proximity effect—2C separation	2×0.28
Proximity effect—1C separation	3×0.28
H attached to a negative group	0.27
Ar–Ar conjugation	0.31
Proximity effects	
(experimental—summation of component fragments)	
–COOH	8×0.28
$–CONH_2$	4×0.28
$–NHCONH_2$	11×0.28

[a] This is only a small part of the scheme reported by Rekker and de Kort (1979).
[b] The symbol in brackets denotes aliphatic (Al) or aromatic (Ar).

shown in Table 9.3. In both cases the procedure for the calculation of a log *P* value is the same; a compound is broken down into the appropriate fragments, correction factors are identified, and the fragments and correction factor values are summed up. Figure 9.1 illustrates the process for three different compounds where it can be seen that both methods can give quite comparable results which are in good agreement with the experimental values. This, of course, is not always the case; for some types of compounds, as shown in the figure, the Rekker method may give better estimates than the Hansch and Leo approach, and vice versa for other sets of compounds. A comparison of the approaches concludes, perhaps unsurprisingly, that neither can be said to be 'best' (Mayer *et al.* 1982) and it is prudent to always compare predictions with measured values whenever possible.

In these two examples, the knowledge base or rule base of the expert system consists of the fragment values and correction factors, along with any associated rules for breaking down a compound into appropriate fragments. Although the numerical values for the fragments and factors have been derived from experimental data (log *P* measurements) it has

Table 9.3. Some fragment values and correction factors for the Hansch and Leo log P prediction system[a] (from Hansch and Leo 1979, with their kind permission)

	Value[b]		
Fragment	f	f^{ϕ}	$f^{\phi\phi}$
Br	0.20	1.09	
Cl	0.06	0.94	
NO_2	−1.16	−0.03	
OH	−1.64	−0.44	
COOH	−1.11	−0.03	
$CONH_2$	−2.18	−1.26	
-O-	−1.82	−0.61	0.53
-NH-	−2.15	−1.03	−0.09
-CONH-	−2.71	−1.81	−1.06
$-CO_2-$	−1.49	−0.56	−0.09

Correction factors	
Type	Value
Normal double bond	−0.55
Conjugate to ϕ	−0.42
Chain single bond	−0.12 (proportional to length)
Intramolecular H-bond	0.60 (for nitrogen)
Intramolecular H-bond	1.0 (for oxygen)

[a] From the scheme (Tables IV-1a and IV-1b reported by Hansch and Leo 1979).
[b] Value given for aliphatic (f), aromatic (f^{ϕ}), and for the fragment between two aromatic systems ($f^{\phi\phi}$).

required a human expert to create the overall calculation scheme. This can be seen particularly clearly for the Hansch and Leo system which, because of the small number of fragments in the scheme, requires a variety of different types of correction factors to account for the way that different fragments influence one another. Devising the correction factors and the rules for their application has required the greatest contribution of human expertise in these two systems. One major problem in the application of any expert system to chemical structures is the question of how to break down compounds into fragments that will be recognized by the system. The problem lies not so much in the process of creating fragments but in deciding which are the 'correct' fragments. Although the rules for creating fragments are a necessary part of such expert systems, it is possible, particularly for large molecules, to create different sets of fragments which still conform to the rules. If the expert system has been carefully created, these different sets of fragments (and correction factors) may yield the same answer, but it is disconcerting to find two (or more) ways to do the same job. This was a particular problem when the schemes were first

$HO—CH_2—COOH$ Measured log $P = -1.11$

Rekker:

$$\text{Log } P = f_{OH,al} + f_{CH_2} + f_{COOH,\,al} + 3CM$$
$$= -1.022 \qquad \Delta = 0.09$$

Hansch and Leo:

$$\text{Log } P = f_{OH} + f_{CH_2} + f_{COOH} + (2\text{-}1)\,F_b + F_{P1}$$
$$= -1.06 \qquad \Delta = 0.05$$

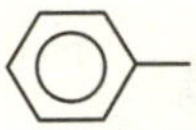

C_6H_5—O—$CH_2\,CH_2\,OH$ Measured log $P = 1.16$

Rekker:

$$\text{Log } P = f_{C_6H_5} + f_{O,or} + f_{CH_2} + f_{OH,\,al} + 2CM$$
$$= -1.55 \qquad \Delta = 0.39$$

Hansch and Leo:

$$\text{Log } P = f_{C_6H_5} + f_O{}^{\phi} + 2f_{CH_2} + f_{OH} + (4\text{-}1)\,F_b + F_{P2}$$
$$= -1.20 \qquad \Delta = 0.04$$

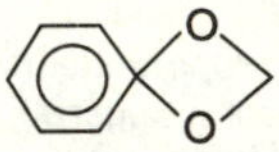

(1,3-benzodioxole) Measured log $P = 2.08$

Rekker:

$$\text{Log } P = f_{C_6H_4} + 2f_{O,or} + f_{CH_2} + 3CM$$
$$= 2.17 \qquad \Delta = 0.09$$

Hansch and Leo:

$$\text{Log } P = 4f_{CH}{}^{\phi} + 2f_C{}^{\phi} + 2f_O{}^{\phi} + f_{CH_2} + 3\,F_b + F_{P1} + F_{P2}{}^{\phi}$$
$$= -1.34 \qquad \Delta = -0.74$$

Fig. 9.1. A comparison of the Rekker and the Hansch and Leo log *P* calculation for three molecules (from Mayer *et al.* 1982, with permission of the *European Journal of Medicinal Chemistry*).

created and all calculations were carried out manually, but more recently they have been implemented in computer systems. The Hansch and Leo system is probably the most widely used, and is available commercially from Daylight Chemical Information Systems (see Software appendix). Of course, a computer implementation of a chemical expert system requires some means by which chemical structure information can be passed to the computer program. Unless the particular expert system simply requires a molecular formula (or some other atom count), this means that it is necessary to provide two- or three-dimensional information to the program. There are a variety of ways in which two- and three-dimensional chemical information can be stored and processed by computers; the input

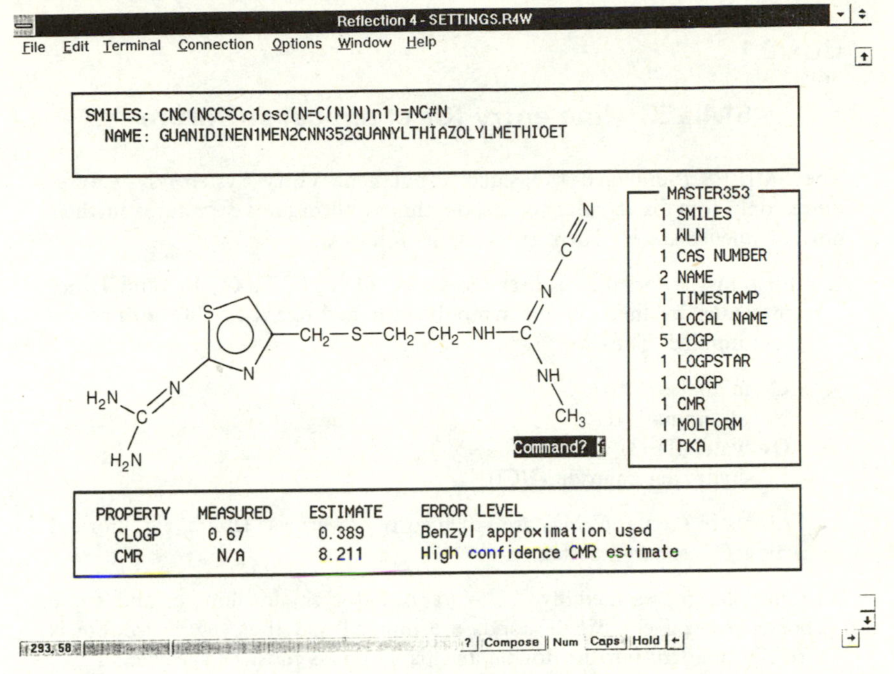

Fig. 9.2. Computer screen for input of the compound tiotidine to the program UDRIVE (with permission of Daylight Chemical Information Systems Inc.).

system used by the Hansch and Leo expert system of log P calculation (CLOGP) is known as SMILES (Simplified Molecular Input Line Entry System). The SMILES coding scheme is so elegantly simple and easy to learn (see Box 9.1) that it has become used as an input system for several other chemical calculation programs. Figure 9.2 shows the input screen for one means of access to the CLOGP program* for the compound tiotidine, an H_2 receptor antagonist used for the control of gastric acid secretion. The SMILES string CNC(NCCSCc1csc(N=C(N)N)n1)=NC#N is shown at the top of the screen with a two-dimensional representation of the compound in the box. A calculated value for log *P* (0.389) and molar refractivity (8.211) appear at the bottom of the screen and, in this case, a measured log *P* value (0.67) which has been retrieved from the Pomona College Medicinal Chemistry Database. The Daylight software provides access to a variety of chemical databases in addition to the CLOGP and CMR calculation routines (see Software appendix). The box on the right-hand side of Fig. 9.2 reports a summary of the types of data which are

* This is the program UDRIVE which provides access to CLOGP and CMR calculation algorithms, as well as the database routine THOR running on DEC VAX computers. See Software appendix for more details.

Box 9.1

SMILES—line entry for chemical structure

The SMILES (Simplified Molecular Input Line Entry System) system requires only four basic rules to encode almost all organic structures in their normal valence states. These rules are as follows.

1. Atoms in the 'organic' subset (B, C, N, O, P, S, F, Cl, Br, and I) are represented by their atomic symbols with hydrogens (to fill normal valency) implied. Thus,

   C—methane (CH_4)
   N—ammonia (NH_3)
   O—water (H_2O)
   Cl—hydrogen chloride (HCl)

   Atoms in aromatic rings are specified by lower case letters, e.g., normal carbon C, aromatic carbon c.

2. Bonds are represented by –, =, and # for single, double, and triple bonds respectively. Single bonds are implied and thus the – symbol is usually omitted, but the double and triple bonds must be specified. Thus,

   CC ethane (CH_3CH_3)
   CCO ethanol (CH_3CH_2OH)
   C=C ethylene ($CH_2{=}CH_2$)
   O=C=O carbon dioxide (CO_2)
   C=O formaldehyde (CH_2O)
   C#N hydrogen cyanide (HCN)

3. Branches are specified by enclosure of the branch within brackets, and these brackets can be nested or stacked to indicate further branching.

2-propylamine	$CH_3CH(NH_2)CH_3$	CC(N)C
isobutyric acid	$CH_3CH(CH_3)C(=O)OH$	CC(C)C(=O)O
3-isopropyl-1-hexene	$CH_2CHCH(CH(CH_3)_2)CH_2CH_2CH_3$	C=CC(C(C)C)CCC

4. Cyclic structures are represented by breaking one single or aromatic bond in each ring and numbering the atoms on either side of the bond to indicate it. This is shown for several different rings in the figure. A single atom may have more than one ring closure; different ring closures are indicated by different numbers (the digits 1–9 are allowed and can be reused after closure of that ring bond).

These simple rules allow very rapid encoding of most chemical structures and need only a few simple additions to cope with other atoms, charges, isomers, etc. Specification of atoms not in the 'organic' subset, for example, is coded by use of an atomic symbol within square brackets. An extensive description of the SMILES system is given by Weininger and Weininger (1990).

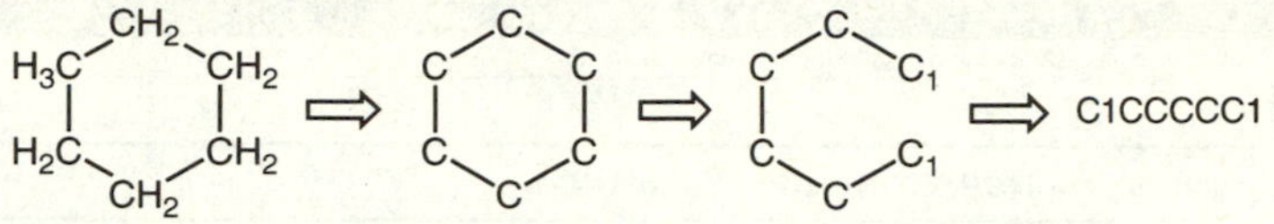

C1CCCCC1

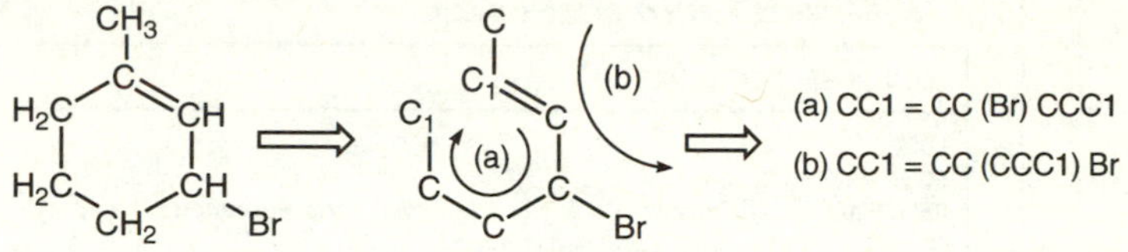

(a) CC1 = CC (Br) CCC1
(b) CC1 = CC (CCC1) Br

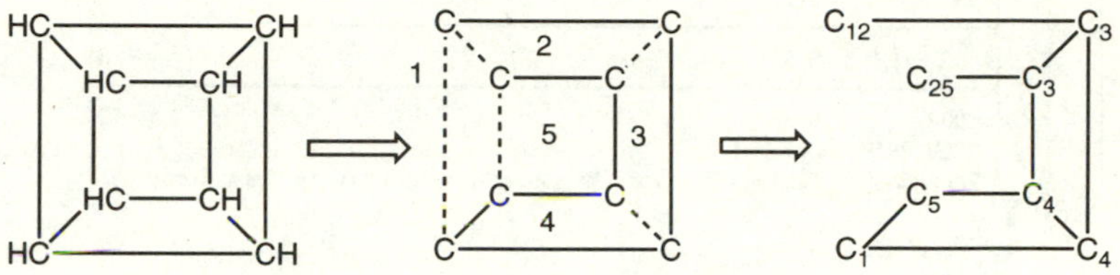

Generation of SMILES for cubane: C12C3C4C1C5C4C3C25

One of the most attractive features of the SMILES structure generation algorithms is that is does not matter where a SMILES string begins—if the coding is correct the corresponding structure will be produced. This is in marked contrast to other linear chemical structure coding schemes where order is important and there are complex rules to decide where to start. The following are all valid SMILES for 6-hydroxy-1,4-hexadiene.

$CH_2{=}CH{-}CH_2{-}CH{=}CH{-}CH_2OH$ C=CCC=CCO
C(C=C)C=CCO
OCC=CCC=C

Although the ordering of coding of a SMILES string does not matter for the input of structures, it does have an effect on the efficiency of storing and subsequent searching of a collection of compounds. The order of coding of structures also has implications for the generation of rules for chemical expert systems which, for example, predict three dimensional chemical structure. By application of a set of ordering rules, it is possible to produce a unique SMILES string for any given structure. This is achieved in the Daylight software by the use of two separate algorithms.

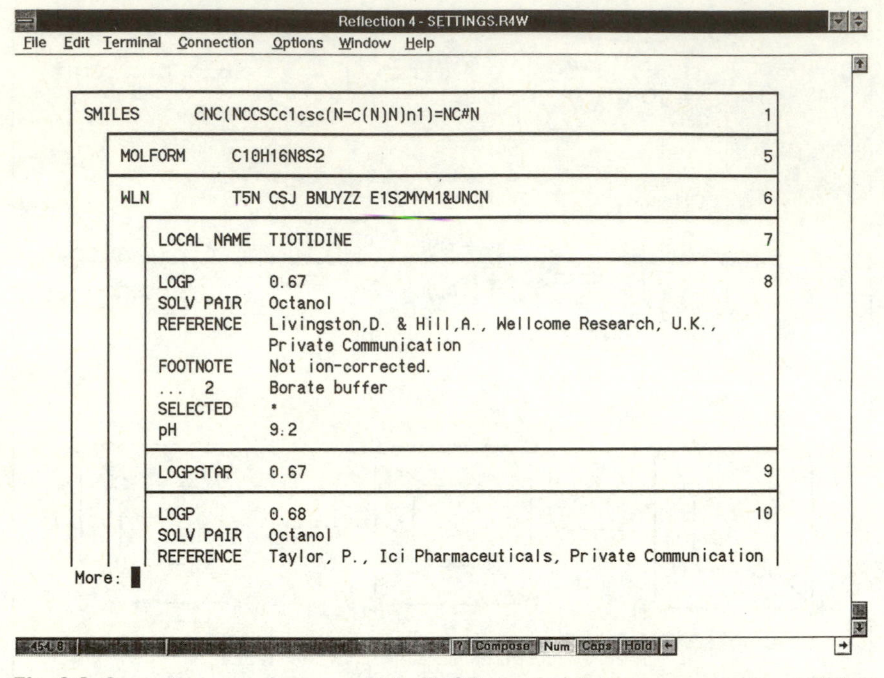

Fig. 9.3. Computer screen for access of the THOR chemical database program from UDRIVE (with permission of Daylight Chemical Information Systems Inc.).

held in the database which the program is currently connected to, in this case the Pomona College Master 353 data collection. Figure 9.3 shows how the data for this compound may be displayed from the UDRIVE menu by accessing the THOR chemical database program. The database page contains the SMILES string for tiotidine as the root of the data tree, the molecular formula and a WLN string,* a local name for the compound, and then experimental values for log P and pK_a in some cases. The measured value of 0.67 has been selected by the database constructors as a 'best' value, called a log P^*. The Pomona College database stores partition coefficient values for octanol/water and other solvent systems (in 1993, 50,000 measured log P values and 9,000 pK_a values for 26,000 compounds). Details of the fragments and correction factors used in the calculation of log P can be obtained from the program as shown in Fig. 9.4. In this case the calculation made use of one approximated fragment as shown by the comments alongside the fragments and factors; the degree of certainty can be seen by comparison of the predicted with the measured value (Δ log $P = -0.28$).

* WLN (Wiswesser Line Notation) is another line notation system for chemical structures which has seen widespread use.

Reflection 4 - SETTINGS.R4W

File Edit Terminal Connection Options Window Help

Class	Type	Log(P) Contribution Description	Comment	Value
FRAGMENT	# 1	cyanoguanidine	MEASURED	-2.130
FRAGMENT	# 2	Sulfide	APPROX.	-0.590
FRAGMENT	# 3	Thiophenyl	MEASURED	0.360
FRAGMENT	# 4	N-guanidyl	MEASURED	-1.400
FRAGMENT	# 5	Aromatic nitrogen (TYPE 2)	MEASURED	-1.120
ISOLATING	CARBON	4 Aliphatic isolating carbon(s)		0.780
ISOLATING	CARBON	3 Aromatic isolating carbon(s)		0.390
EXFRAGMENT	HYDROG	10 Hydrogen(s) on isolating carbons		2.270
EXFRAGMENT	BONDS	5 chain and 0 alicyclic (net)		-0.600
PROXIMITY	Y-CC-Y	1 pairs over bond 6- 5 (AvWt=-.260)		0.707
ELECTRONIC	SIGRHO	2 Potential interactions; 2.00 used	WithinRing	1.722
RESULT	v3.4	Benzyl approximation used	CLOGP	0.389

Nomore:

545 10 ? Compose Num Caps Hold

Fig. 9.4. Details of the calculation of log *P* for tiotidine (computer screen from UDRIVE) (with permission of Daylight Chemical Information Systems Inc.).

9.2.2 Toxicity prediction

The structural alert model for carcinogenicity first proposed by Ashby (1985) and later modified by Tennant and Ashby (1991) is a good example of a manual expert system in chemistry. This scheme was created by the recognition of common substructures which occur in compounds which have shown positive in an *in vitro* test for mutagenicity, the well-known AMES salmonella test. Putting together these substructures has allowed the creation of a 'supermolecule' as shown in Fig. 9.5. Prediction of the likelihood of mutagenicity for any new compound is achieved by simple comparison of the new structure with the alerts present in the supermolecule; common sense suggests that the greater the number of alerts, then the higher the likelihood of mutagenicity. The structural alert system has been shown to be successful for the prediction of mutagenicity; for example, in a test of 301 chemicals (Ashby and Tennant 1991) almost 80 per cent of alerting compounds were mutagenic compared with 30 per cent of non-alerting compounds. Unfortunately, compounds may be carcinogenic due to mechanisms other than mutagenicity (thought to be caused by reaction with a nucleophilic site in DNA). The correlation between mutagenicity and carcinogenicity (measured in two rodent species) was

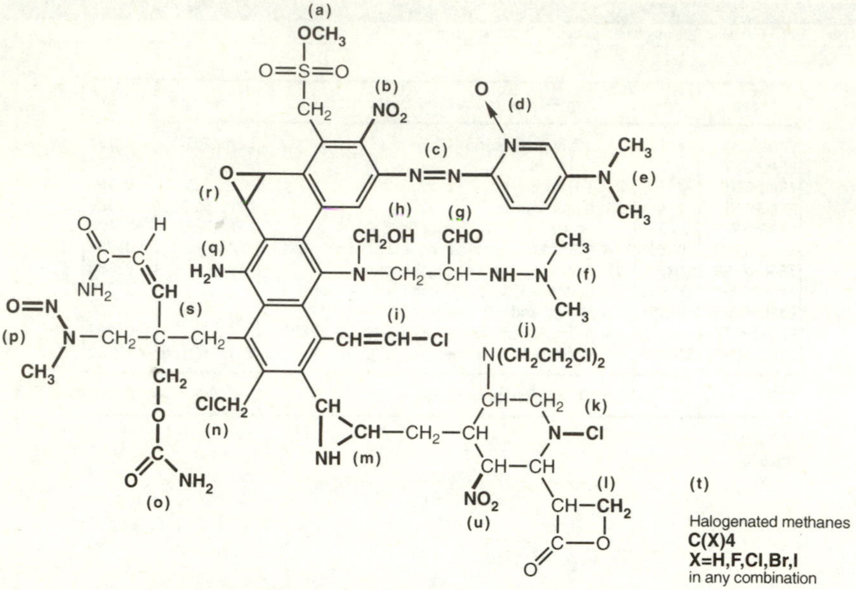

Fig. 9.5. A supermolecule for the assessment of structural alerts for toxicity (from Tennant and Ashby 1991, with permission of Elsevier Science).

low, suggesting that structural alerts are useful but non-definitive indicators of potential carcinogenic activity (Ashby and Tennant 1991). This is a nice example of the strengths and limitations of expert systems: if the experts' knowledge is well coded (in the rules) and used correctly, the system will make good predictions. The expert system, however, can only predict what it really knows about; if a compound is carcinogenic because it is mutagenic then all is well.

The DEREK system (Deductive Estimation of Risk from Existing Knowledge) is a computer-based expert system for the prediction of toxicity (Sanderson and Earnshaw 1991). This program uses the LHASA synthesis planning program (see Section 9.2.3) as its foundation for the input of chemical structures and the processing of chemical substructures. DEREK makes predictions of toxicity by the recognition of toxic fragments, toxicophores, defined by the rules present in a rule base created by human experts. At present, the rule base is being expanded by a collaborative effort involving pharmaceutical, agrochemical, and other chemical companies, as well as government organizations. The collaborative exercise involves a committee which considers any new rules that are presented for inclusion in the rule base. Rules are written in the PATRAN language of LHASA, and when a rule is activated, due to the presence of a toxic fragment, a different language, CHMTRN, is used to consider the rest of the structure and the environment of the toxicophore. Toxicity is not determined simply by the presence of toxic fragments, but also by other features in the compound which may modify the behaviour of toxicophores. Figure 9.6 gives an example of the display of a DEREK answer for a query compound (part a) and display of the notes written by the rule writer for the rule which has been activated for this compound (part b). Prediction of the potential toxicity of compounds is of considerable appeal to most areas of the chemical industry, particularly if this reduces the need for animal testing. Unfortunately there is a major complication in the prediction of toxicity in the form of metabolism. Indeed, there are a number of therapeutic compounds which rely on metabolism to produce their active components. Currently, metabolic processes are dealt with in the DEREK system by explicit statements in the rule base (Langowski 1993) and for a method such as the identification of structural

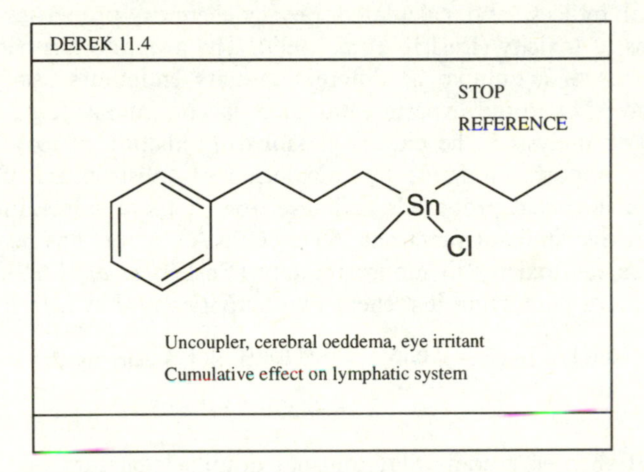

DEREK 11.4 Reference display

Rule: 17
Risk: Uncoupler, cerebral oedema, eye irritant
File: DEREK1

Trialkyl tin compounds

R3Sn-X or R4Sn which can be metabolically dealkylated

If any of the alkyl groups are large, then cumulative effects on the lymphatic system are possible

Drop pen to continue

Fig. 9.6. An example of the output from the DEREK program (from Judson 1992, with permission of the Society of Chemical Industry).

alerts, the alerting fragments may well have been chosen because human metabolism leads to toxic structures. In principle, the likely routes of metabolism for a given compound in a particular species can be predicted by an expert system, and programs exist which aim to do just that (see Section 9.4). The production of a generally applicable expert system for toxicity prediction is likely to require an expert system for metabolism predictions as well as perhaps some means of assessing distribution and elimination.

Other methods have been reported for the estimation of toxicity and these are often called 'expert systems' although the definition of an expert system used here, in which humans define the rules, would not label them as such. The TOPKAT program (Toxicity Prediction by Komputer Assisted Technology) uses a combination of substructural descriptors, topological indices, and calculated physicochemical properties to make predictions of toxicity (Enslein *et al.* 1990). The program has been trained on databases of a number of different toxicity endpoints using multiple linear regression if the experimental data is continuous (e.g., LD_{50}) or discriminant analysis if the data is classified. Prediction of toxicity for an unknown compound is made by calculation of substructural descriptors and physicochemical properties and insertion of these values into the regression or discriminant functions. One TOPKAT model has been created which links rat toxicity to mouse toxicity (Enslein *et al.* 1989). A 'rudimentary' model involving just these two activities is shown in eqn (9.1).

$$\log(1/C)_{\text{RAT}} = 0.636 + 1.694 \log(\log(1/C)_{\text{MOUSE}} + 1) \qquad (9.1)$$

$$N = 160 \quad R^2 = 0.399 \quad \text{SE} = 0.41.$$

This equation uses a somewhat unusual double logarithm of the mouse toxicity data, to make this variable conform better to a normal distribution. A better predictive equation for rat toxicity was obtained by combining structural descriptors with the mouse toxicity data as shown in Table 9.4.

Input of chemical structure to the TOPKAT program is by means of SMILES; the program recognizes fragments and properties which correspond to the regression equation or discriminant function which is associated with the selected module (particular toxic endpoint) and calculates a value for the endpoint. The parts of the structure which are used in the calculation are shown graphically, as can be seen in the top part of Fig. 9.7. The lower part of the figure gives details of the calculation of toxicity (LC_{50} for Fathead Minnow) for the compound, which in this case gives a value of 4.029 for LC_{50} (equivalent to 13.3 mg/l). The program also recognizes features in a molecule that are not present in the predictive model, and will issue a warning if the compound has not been well 'covered'. Figure 9.8 shows two output screens for the prediction of toxicity of a herbicide to

Table 9.4. Regression model for the prediction of rat LD_{50} from mouse LD_{50} and structural descriptors (from Enslein *et al.* 1989, with permission of Princeton Scientific Publishers)

WLN Key #	Variable description	Coefficient	*F*
	$\log(\log(1/C)_{\text{MOUSE}}+1)$	1.654	227.3
144	Two heterocyclic rings	−0.361	12.4
	Molecular weight	0.000432	11.9
10	One sulphur atom	0.125	11.4
39	One − C = O (chain fragment)	−0.0965	11.1
58	One $-NH_2$ group	−0.0990	7.76
	$(\log P)^2$	−0.00833	5.71
107	One heteroatom in more than one ring	0.191	7.00
37	One − OH group (chain fragment)	−0.0605	5.25
135	Two ring systems (not benzene)	0.186	5.15
67	One − C = O group (substituent)	−0.0856	4.56
	Constant	0.628	

$R^2 = 0.793$ SE = 0.33 $N = 160$

Daphnia magna. The top part of the figure shows substructural features that are not in the predictive equation but which are common to the compounds in the database from which the equation was developed. This screen gives a warning that the prediction may be unreliable. The second part of the figure shows one of the screens which gives details of the calculation. The recognition of incomplete coverage is particularly important for any method that relies on the use of certain substructures or physicochemical features for prediction. It is essential to know how much of the molecule was recognized and used in the calculation in order to be able to make some sort of judgement of the reliability of prediction.

The CASETOX program uses a somewhat similar approach to TOPKAT but rather than using predetermined structural fragments, CASETOX generates all possible fragments for the compounds in a database (Klopman 1985). Like TOPKAT, the CASETOX system has been trained on databases for a number of toxic endpoints. Perhaps one of the biggest problems with any of these toxicity prediction systems lies in the quality and suitability of the available databases. Since these databases often collect together results from different laboratories, the question of consistency obviously arises. Suitability is probably an even more important matter. If the compound for which prediction is required is a potential pharmaceutical, then the best database for prediction will contain mostly pharmaceuticals, preferably with some degree of structural similarity. Such databases are not yet generally available.

Finally, how well do such prediction systems work? In 1990, a challenge was issued to interested parties to make predictions for 44 compounds that were then being tested for rodent carcinogenicity in bioassays by the

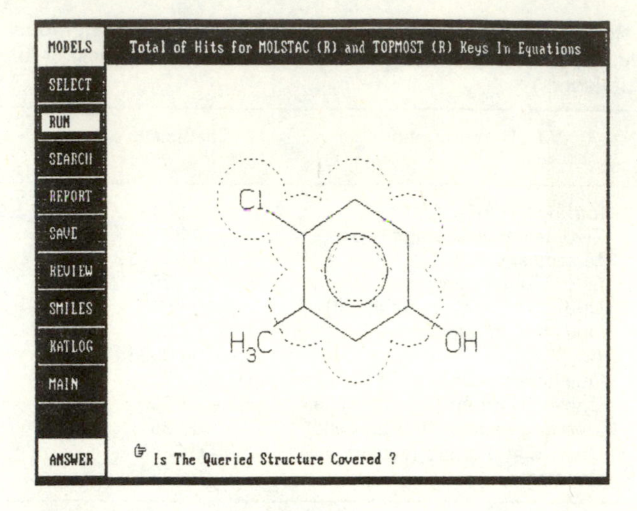

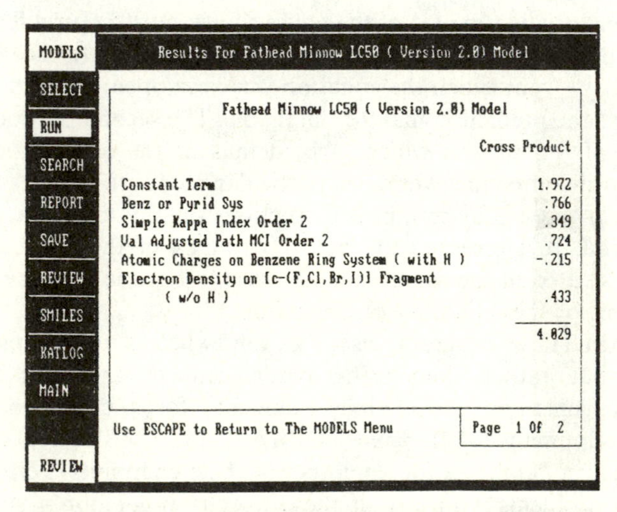

Fig. 9.7. Computer screens from the TOPKAT program showing a prediction of toxicity for 4-chloro-*m*-cresol against Fathead Minnow (figures kindly supplied by Health Designs Inc.).

National Toxicology Program (NTP) in America. The results of the predictions were discussed at a workshop in 1993 and are shown in Table 9.5 (Hileman 1993).* The four computer programs, MULTICASE, DEREK, TOPKAT, and COMPACT, were not particularly impressive in their predictions since their best result was 62 per cent correct and the requested prediction was YES/NO, which might be expected to be 50 per cent right

* At the time of the meeting, results had been obtained for 40 of the 44 compounds.

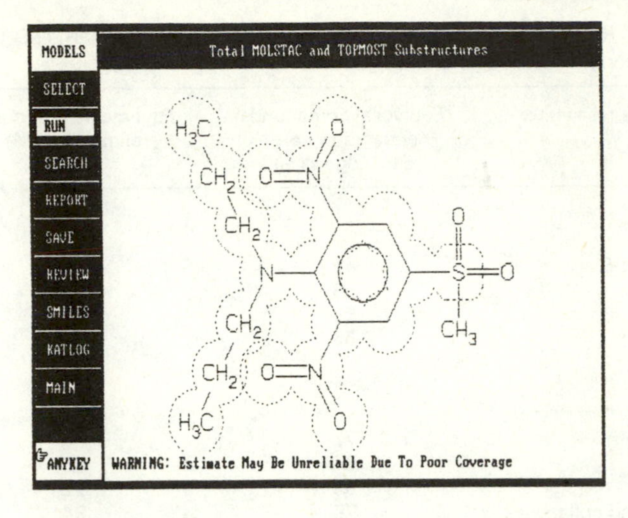

MODELS | Results For Daphnia Magna 48 Hr Ec50 Model

SELECT · RUN · SEARCH · REPORT · SAVE · REVIEW · SMILES · KATLOG · MAIN · REVIEW

Daphnia Magna 48 Hr Ec50 Model

	Cross Product
Longest At Chain in Rg Mol	.771
Benzene Ring System	.722
Val Adjusted Path MCI Order 1	2.289
Sigma Charges on Benzene Ring System (w/o H)	.559
Sigma Charges on Nitrogen Atom in [N=O] Fragment (w/o H)	-.773
Constant Term	2.287
	5.774

Use ESCAPE to Return to The MODELS Menu | Page 1 Of 2

Fig. 9.8. Computer screens from the TOPKAT program showing a prediction of toxicity for nitralin against *Daphnia magna* (figures kindly supplied by Health Designs Inc.).

by chance, given an even distribution of carcinogens in the set. The results from the human groups or individuals are at least as good or considerably better, but in fairness it should be pointed out that in some cases these predictions made use of more information than was available to the programs. One other consideration should be borne in mind when comparing prediction results between computer systems, or between computers and humans, and that is the overall number of predictions made. The

Table 9.5. Results of carcinogenicity prediction (from Hileman 1993, copyright (1993) American Chemical Society)

Program or researcher	Equivocal compounds equated to non-carcinogen (%)	Equivocal compounds eliminated (%)
Programs		
MULTICASE	49	55
DEREK	59	62
TOPKAT	58	58
COMPACT	56	62
Humans		
Rash	71	68
Bakale	64	65
Benigni	63	72
Tennant and colleagues	75	84

human experts generally made a prediction for every compound, whereas different computer systems omitted different numbers of compounds because their 'rules' could not cope with certain structures or sub-structures. There is no doubt that the performance of these computer systems will improve as they continue to be developed.

9.2.3 Reaction routes and chemical structure

The LHASA system (Logic and Heuristics Applied to Synthetic Analysis) has already been mentioned as the basis of the DEREK toxicity prediction program. The LHASA program was originated as OCSS (Organic Chemical Simulation of Synthesis) by E. J. Corey at Harvard University (Corey *et al.* 1985) and is being further developed both in the United States and UK (LHASA UK, University of Leeds). This program contains a large database of organic reactions, also known as a knowledge base, and a set of rules (heuristics) that dictate how the reactivity of particular functional groups or fragments (retrons) is affected by other parts of the molecule, reaction conditions, etc. The aim of the program is to suggest possible synthetic routes to a given target molecule from a particular set of starting materials. This system actually starts at the target structure and breaks this down into simpler materials until eventually reaching the starting compounds, thus the process is retrosynthetic. An example of one of the steps given by the program for the synthesis of a prostaglandin precursor is shown in Fig. 9.9. When setting up the synthesis query, the user can select one of five different synthetic strategies, including a

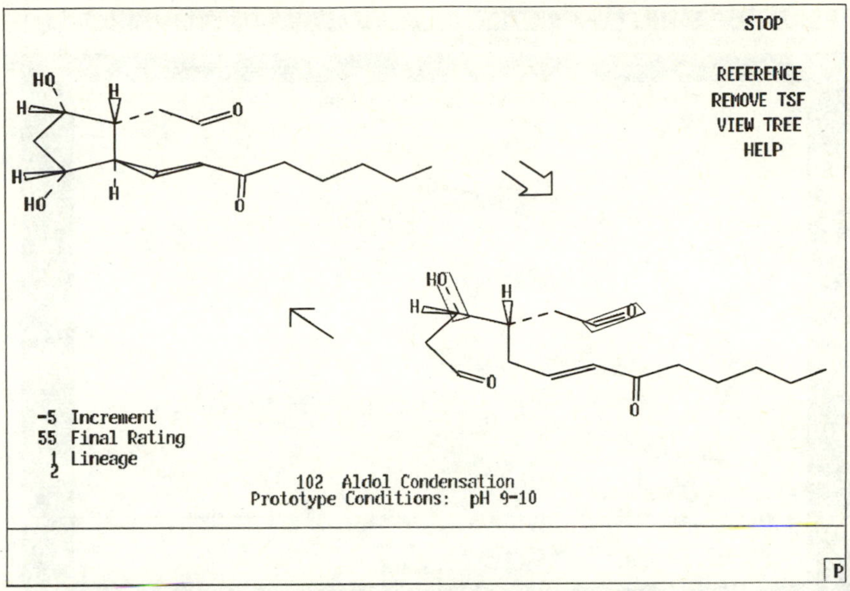

Fig. 9.9. Computer screen from the LHASA program showing a reaction step (with permission of LHASA UK Ltd).

stereochemical option, and there is also an option for the program to make its own suggestions for strategies and tactics. Synthesis of a relatively complex molecule may proceed from simple starting materials by a great many different routes; Corey (1991), for example, shows a retrosynthetic analysis of aphidicolin produced by LHASA which contained over 300 suggested intermediates. Figure 9.10 shows an early stage in the synthesis planning of the prostaglandin precursor shown in Fig. 9.9. Each number (node) on the display represents a different compound—Fig. 9.9, for example, involved nodes 1 and 2.

Of course synthesis-planning expert systems can operate in the opposite direction, from starting materials to products, and the CAMEO program (Computer Assisted Mechanistic Evaluation of Organic reactions) is an example of this (Metivier *et al.* 1987). A problem with the operation of a synthesis-planning system in the forward direction, particularly if several steps are required, is the potentially large number of synthetic alternatives that must be considered.* Each intermediate molecule may undergo a number of transformations and thus the target may be reached by a number of different routes. A successful expert system not only has to work out the feasible routes, but also has to assign some likelihood of success, or degree of difficulty, to the individual routes. Figure 9.11 shows

* The same is also true of retrosynthetic systems.

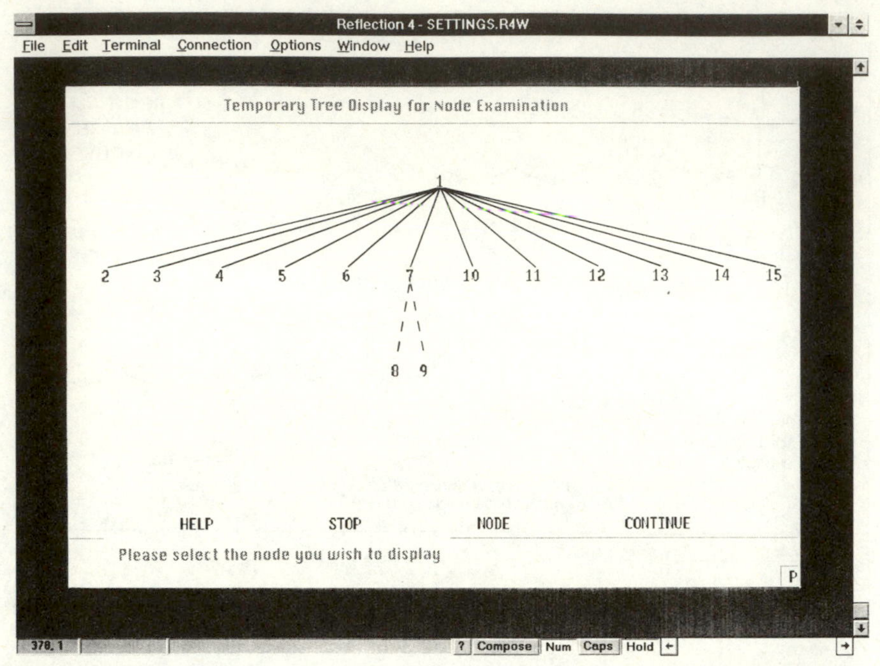

Fig. 9.10. Computer screen from the LHASA program showing an intermediate stage in the breakdown of the prostaglandin precursor shown in Fig. 9.9 (with permission of LHASA UK Ltd).

an example of an output screen from the CAMEO program for a reaction predicted using the acidic/electrophilic mechanistic module. This screen shows that this compound is one of the possible products, the smiling face symbol indicates that the program predicts that this will be a major product, and there is even a calculated ΔH for the reaction. The CAMEO program has a quite comprehensive set of options for controlling the conditions under which reaction predictions will be made (Fig. 9.12). The user can choose one of eight different mechanistic modules, temperature ranges may be set, and there are menus for the choice of reagents and solvents. Another view of the CAMEO sytem is that it is a reaction evaluaton program. It may be used to verify suggestions, made by other programs or methods (or even individual chemists) concerning particular reactions within a sequence.

This account of the LHASA and CAMEO programs has been necessarily very limited; readers interested in further details should contact the suppliers of the software as listed in the Software appendix. Another approach to the problem of synthesis planning is to provide literature references to model reactions, thus allowing the chemist user to make his or her own assessment of feasibility, using his or her own expert system! A number of reaction database systems are in common use (e.g., REACCS,

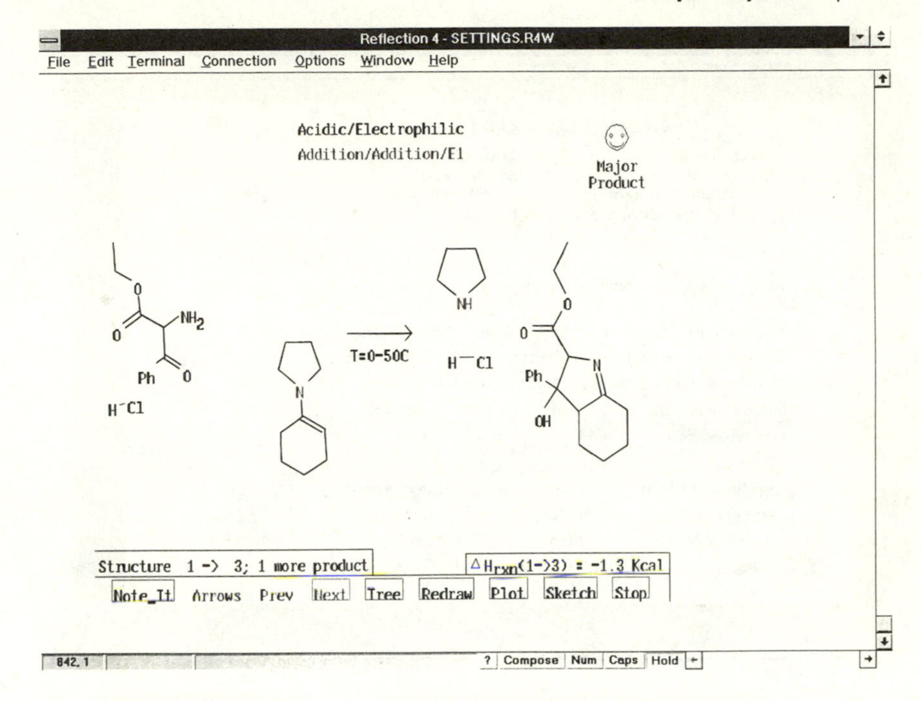

Fig. 9.11. Output screen from the reaction prediction program CAMEO (with permission of LHASA UK Ltd).

SYNLIB, see Software appendix) and a combination of synthesis-planning software with literature reaction retrieval software is a powerful tool for synthetic chemistry.

The prediction of three-dimensional chemical structure from a list of atoms in a molecule and their connectivity is a good example of a chemical problem that may be solved by an expert system. We have already seen (Fig. 9.2) how the SMILES interpreter can construct a two-dimensional representation of a structure from its one-dimensional representation as a SMILES string. The CONCORD program (CONnection table to CoORDinates) takes a SMILES string and, very rapidly, produces a three-dimensional model of an input molecule. This system is a hybrid between an expert system and a molecular mechanics program, molecular mechanics being the method by which molecular structures are 'minimized' in most molecular modelling systems. The procedure operates as follows.

1. The SMILES input string is checked for syntax errors and is made 'unique' if it is not already (see Box 9.1).
2. The atom symbols in the SMILES string are numbered sequentially and a connection table is constructed which indicates which atoms are bonded to each other.

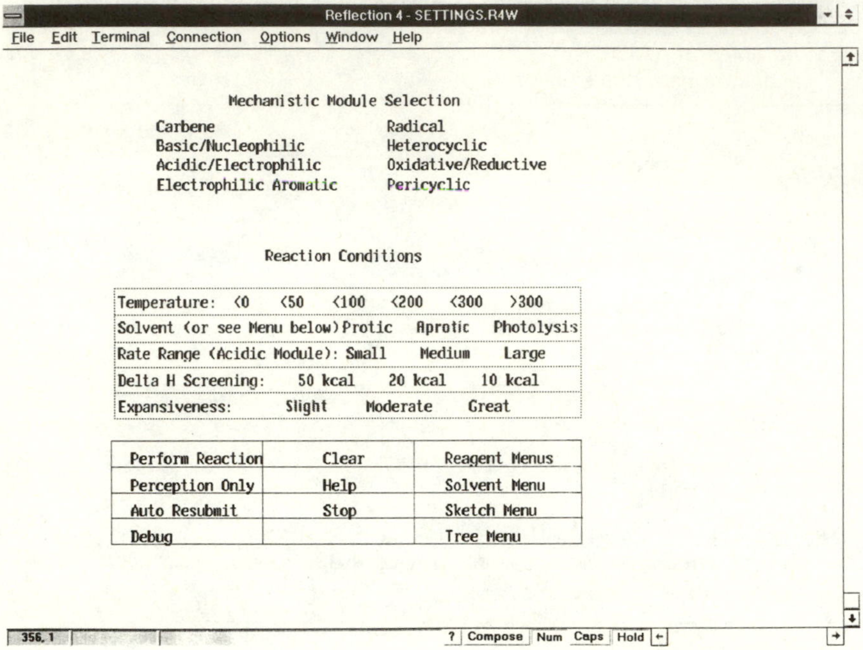

Reflection 4 - SETTINGS.R4W

File Edit Terminal Connection Options Window Help

Mechanistic Module Selection

Carbene	Radical
Basic/Nucleophilic	Heterocyclic
Acidic/Electrophilic	Oxidative/Reductive
Electrophilic Aromatic	Pericyclic

Reaction Conditions

Temperature: <0 <50 <100 <200 <300 >300

Solvent (or see Menu below) Protic Aprotic Photolysis

Rate Range (Acidic Module): Small Medium Large

Delta H Screening: 50 kcal 20 kcal 10 kcal

Expansiveness: Slight Moderate Great

Perform Reaction	Clear	Reagent Menus
Perception Only	Help	Solvent Menu
Auto Resubmit	Stop	Sketch Menu
Debug		Tree Menu

356, 1 ? Compose Num Caps Hold

Fig. 9.12. Output screen from the CAMEO program showing the choice of mechanistic modules and reaction conditions for prediction (with permission of LHASA UK Ltd).

3. A bond type table (i.e., single, double, aromatic, etc.) is constructed from the connection table and a ring table is constructed from the connection table.
4. The connection table and bond table are used to assign the number of hydrogens to be attached to each atom and other chemical validity checks are carried out.

A set of rules is used to assign reasonable three-dimensional structures to various features in the molecule, e.g., rings, and to certain bond angles and torsion angles. At this point the three-dimensional structure will be a good approximation for most parts of the molecule but some bond angles and torsion angles could be adjusted to minimize unfavourable steric interactions (at the expense of introducing extra energy into the system in the form of bond angle or torsion angle strain). This requires optimization of an expression for energy such as that shown in eqn (9.2).

$$E = \sum_{i=1}^{N} k_i^l (b_{0,i} - b_i) + \sum_{i=1}^{N} k_i^\theta (\Theta_{0,1} - \Theta_i) + \ldots \tag{9.2}$$

Table 9.6. Performance of CONCORD on a subset[a] of the Cambridge Structural Database (X-ray structures) and the fine chemical directory (FCD) and modern drug data report (MDDR)[b]. (Personal communication from R. S. Pearlman, College of Pharmacy, the University of Texas at Austin.)

	CSD	FCD and MDDR
Number of compounds	30 712	101 776
Total errors	3 207 (10.4%)	1 531 (1.5%)
made up from		
ring-system	707 (2.3%)	223 (0.2%)
chiral fusions	1 462 (3.4%)	216 (0.2%)
close-contacts	1 038 (3.4%)	1 092 (1.1%)
Close-contact warnings[c]	11 126 (36.2%)	28 080 (27.6%)

[a] Excluding 'unusual' atoms (e.g., As, Se, etc).
[b] Compounds are often chosen for X-ray structure determination because they represent unusual structural classes, hence the relatively high error rate compared with the more 'typical' structures in the FCD and MDDR.
[c] Structures are generated in these cases and may be 'cleaned up' by molecular mechanics or quantum mechanics calculations.

In this very simple expression for the energy of a molecule, the first two terms represent a summation over all the bonds of the compound of the energy contribution due to bond lengths and bond angles. The two constants, k_i^l and k_i^θ, represent the force necessary to distort that particular type of bond (i) from its 'natural', i.e. average, bond length ($b_{0,i}$) or bond angle ($\Theta_{0,i}$). The standard bond lengths and bond angles have been derived from experimental measurements such as X-ray crystallography, and the force constants from spectroscopic data. Equation (9.2) is known as a molecular mechanics force field and can be elaborated to include contributions from torsion angles, steric interactions, charge interactions, and so on. The CONCORD system gives warnings of close interactions and other recognized problems in the final structure and these, of course, can be resolved manually or by structure optimization using another molecular mechanics program or by quantum mechanics. The combination of expert system and 'pseudo-molecular mechanics' generally does a good job, as shown in Table 9.6, and has the distinct advantage that it is very fast in terms of computer time. The input of SMILES strings to CONCORD can be automated and in this way many large corporate databases of (generally) good three-dimensional structures have been generated.

9.3 Neural networks

As was briefly mentioned in the introduction to this chapter, artificial neural networks (ANN) are attempts to mimic biological intelligence systems (brains) by copying some of the structure and functions of the components of these systems. The human brain is constructed of a very large number ($\sim 10^{11}$) of relatively slow and simple processing elements called neurons. The response time of a neuron (i.e., the time between successive signals) is of the order of a tenth to one hundredth of a second. In computing terms this is equivalent to a 'clock speed' of 0.01 to 0.1 kHz, very slow compared with the processor speeds of commonly used personal computers (25, 33, 66, or even 100 MHz). So what is it that makes man so smart? The answer lies in the fact that the brain contains a large number of processing elements which are working all the time; this is parallel processing on a grand scale, and the brain in computing terms is a massively parallel device. The other important feature of biological intelligence is the highly complex 'wiring' which joins the neurons together; a single neuron may be connected to as many as 100,000 other neurons.

So what do these processing elements do? Even a cursory examination of a textbook of neurobiology will show the complexity of the biochemical processes which take place in the brain. Various compounds (neurotransmitters) are involved in the passage of signals between neurons, and the functions of the neurons themselves are regulated by a variety of control processes. Ignoring the complexity of these systems the functions of a neuron can be summarized as follows.

1. The receipt of signals from neurons connected to it. These signals can be excitatory or inhibitory.
2. Summation of the input signals, and processing of the sum to reach a 'firing' threshold.
3. The production of an output signal (firing) as dictated by (2) and transmission of this signal to other connected neurons.

This highly simplified description of how a biological neuron functions may not be a good model for the real thing but it serves as the basis for the construction of ANN. Intelligence in living biological systems appears to reside in the way that neurons are connected together and the 'strength' of these connections. Indeed, the creation of connections and the modification of connection weights is thought to be part of the processes involved in our development, i.e. learning and memory. It may not be clear where in the brain signals arise and which pathways they follow, although for certain regions of the brain, such as the sensory organs, it is more obvious. The eyes, for example, produce nervous signals in response to light and these are passed to the visual cortex. Some preprocessing of the information received by the eyes is carried out by sets of neurons which are organized in particular structures, e.g., layers. It is these three functions of biological neurons and their physical organization and connectivity which forms the basis of the construction of ANN.

ANN, like their biological counterparts, are built up from basic processing units as shown in Fig. 9.13. This artificial neuron receives one or

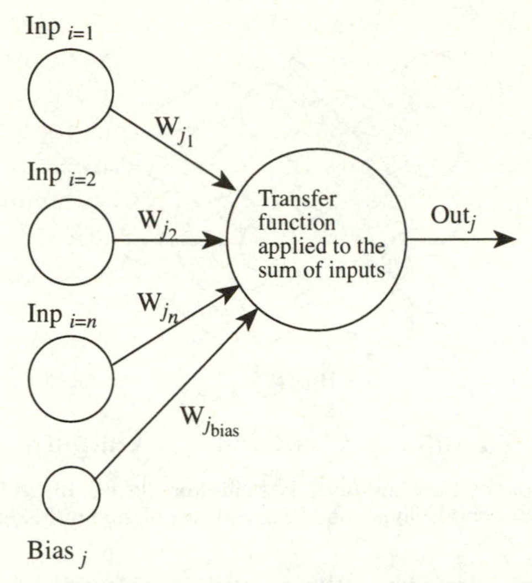

Fig. 9.13. Diagram of an artificial neuron connected to three input neurons and a bias unit (from Manallack and Livingstone 1992, with permission of Birkhauser).

more input signals, applies some kind of transformation function to the summed signal and produces an output signal (equal to the transformation) to be passed on to other neurons. A network usually receives one or more input signals and the input neurons, one for each input signal, behave somewhat differently in that they usually do not do any processing but simply act as distributors to deliver the signal to other neurons in the network. There are many ways in which the neurons in an ANN can be connected together, often referred to as the ANN 'architecture', but one of the most common is in the form of layers as shown in Fig. 9.14. This is known as a 'feed-forward' network since information enters the network at the input layer and is fed forward through the hidden layer(s) until it reaches the output layer. Each neuron in the input layer is connected to every neuron in the hidden layer and each hidden layer neuron in turn is connected to every neuron in the next layer (hidden or output). The strengths or weights of the connections between each pair of neurons are adjustable and it is the adjustment of these connection weights that constitutes training of a network. The neurons, or processing elements, in the hidden layer(s) and output layer apply a non-linear function (transfer function) to their summed inputs such as that shown in eqn (9.3),

$$\text{Output} = \frac{1}{[1 + e^{-s}]} \tag{9.3}$$

where s represents the sum of the inputs to the neuron and Output represents its output signal. The shape of this function is sigmoid as shown in Fig. 9.15 and thus the neuron mimics, to some extent, the way that

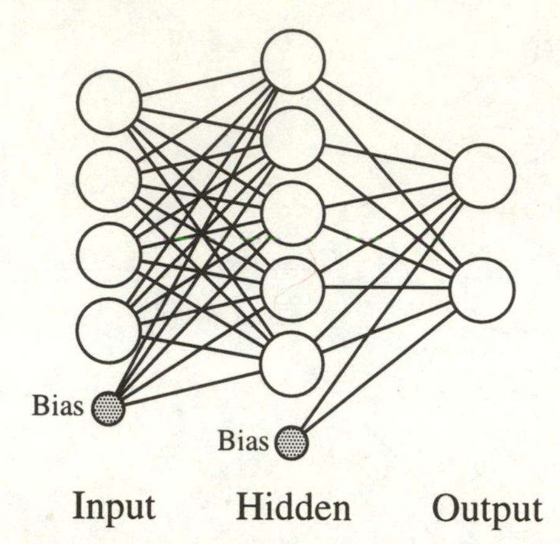

Fig. 9.14. Illustration of how an ANN is built from layers of artificial neurons (from Manallack and Livingstone 1992, with permission of Birkhauser).

biological neurons 'fire' when their input signals exceed some threshold. The use of a function such as that shown in eqn (9.3), or some other non-linear function, allows a network to 'build' non-linear relationships between its inputs and some desired target output.

The bias neurons, one for each layer, represent neurons which produce a constant signal. Their function is to act as shift operators so that the summed inputs for the neurons in the next layer are 'moved', on their transfer function scales, so as to produce signals. Training (adjustment of connection weights) is usually carried out in order to produce a desired target signal or signals at the output layer. A commonly used form of training is known as 'back-propagation of errors' and thus networks such as that shown in Fig. 9.14 sometimes glory under the title of 'back-

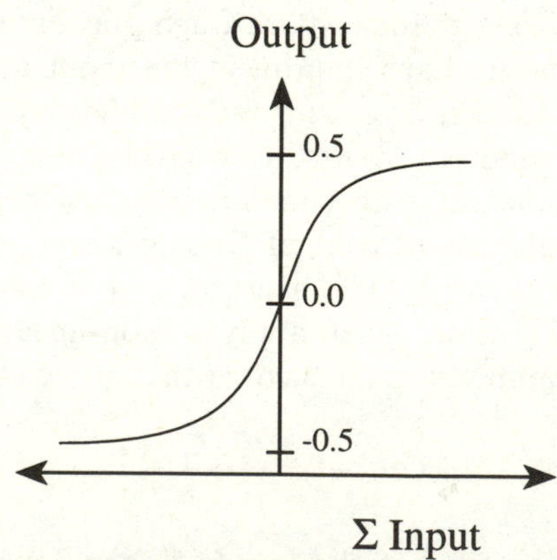

Fig. 9.15. Representation of a commonly used transfer function. Scaled between −0.5 and +0.5, as often used in neural network programs. (from Manallack and Livingstone 1992, with permission of Birkhauser).

propagation feed-forward networks'. There are various ways in which connection weights can be assigned (before training) and networks can be trained. One very common procedure is to assign random values to the connection weights and then to repeatedly pass the training set data through the network, adjusting the weights by back-propagation, until some target error (Δ target – output) is achieved. Of course this does not guarantee that the 'best' solution has been reached and so the network is 'shaken' by applying small perturbations to the weights and then re-training, usually for a set number of passes through the data. This process can be repeated several times, storing the network connection weights and errors at the end of each training cycle, so that the 'best' network can be selected. For a more detailed description of the operation of neural networks see Salt *et al.* (1992), Manallack and Livingstone (1994), and the references contained therein. Reviews on the use of neural networks in chemistry have been published by Zupan and Gasteiger (1991), Burns and Whitesides (1993), Jakus (1992), and Zupan and Gasteiger (1993), and on the use of networks in drug design by Manallack and Livingstone (1994) and Livingstone and Salt (1994). Finally, before moving on to the applications of ANN in data analysis, it is necessary to consider how networks are implemented. ANN are well suited to construction using dedicated computer hardware, particularly when we consider that they are meant to mimic a parallel-computing device. Hardware implementations have the advantage that they can be trained very quickly even when using very large data sets. The disadvantage of constructing networks in hardware, however, is that it is difficult or impossible to change the architecture of the network. Software implementations, although slower to train, are more versatile and are available, both commercially and as 'public-domain software', for a variety of computers (see Software appendix).

9.3.1 Data display using ANN

The physicochemical properties that describe a set of molecules may be used as the input to a neural network and the training target may be some classification (discriminant analysis) or continuous dependent variable (regression analysis) as described in the next section. The training target may also be the values of the input variables themselves and this is the way that a ReNDeR (Reversible Non-linear Dimension Reduction) network operates. A ReNDeR network (Fig. 9.16) consists of an input layer, with one neuron for each descriptor, a smaller hidden layer (encoding), a parameter layer of two or three neurons, another hidden layer (decoding), and an output layer (Livingstone *et al.* 1991). The encoding and decoding hidden layers are of the same size and there are as many output neurons as there are input. Each compound (or sample or object) in a data set is

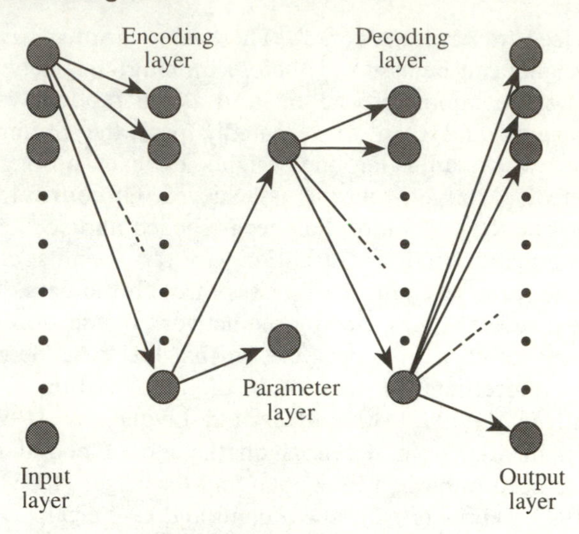

Fig. 9.16. Diagram of a ReNDeR dimension reduction network (from Livingstone *et al.* 1991, copyright 1991 by Elsevier Science Inc.).

presented to the network by feeding in the values of its descriptor variables to the input neurons. The signals from the output neurons are compared to their targets, in this case the value of the input variables, for each compound and the weights in the network are adjusted until the output matches the input. Once training is complete the network has mapped the input onto the output by going through a 'bottleneck' of two (or three) neurons. Each sample in the data set can now be presented to the ReNDeR network in turn and a value will be produced at each of the neurons in the parameter layer. These values will be the summation of the inputs, multiplied by their connection weights, received by the parameter neurons from the neurons in the encoding layer. The parameter layer numbers may be used as x and y (or x, y, and z) coordinates to produce a plot of the samples in the data set. Such a plot will be non-linear because of the network connections and non-linear transfer functions, and provides an alternative method for the low-dimensional display of a high-dimensional data set. An example of this type of display is shown in Fig. 9.17 for a set of 16 analogues of antimycin-a_1 described by 23 calculated physicochemical parameters (these are the same compounds reported in Table 3.1). The active compounds, shown as filled squares, are grouped quite tightly together and thus the plot might be expected to identify new compounds which will be active. The inactive compounds, along with three intermediates, lie in a quite different region of space in this display. For comparison, a non-linear map of this data set is shown in Fig. 9.18 where once again it can be seen that the active compounds are grouped together, although there are some inactives nearby. The non-linear map in

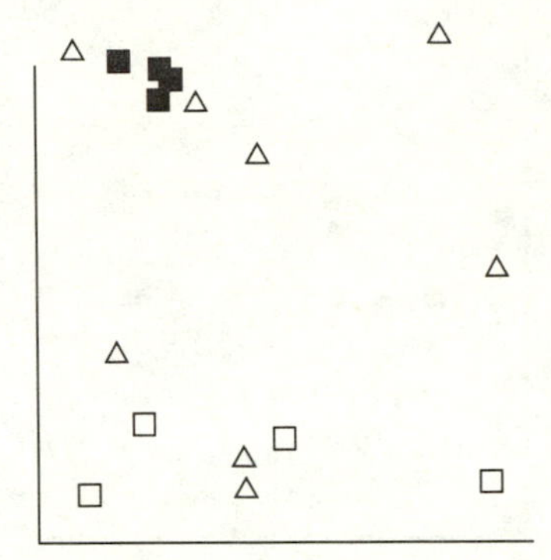

Fig. 9.17. ReNDeR plot of a set of active (■), intermediate (Δ), and inactive (□) compounds described by 23 properties (from Livingstone *et al.* 1991, copyright 1991 by Elsevier Science Inc.).

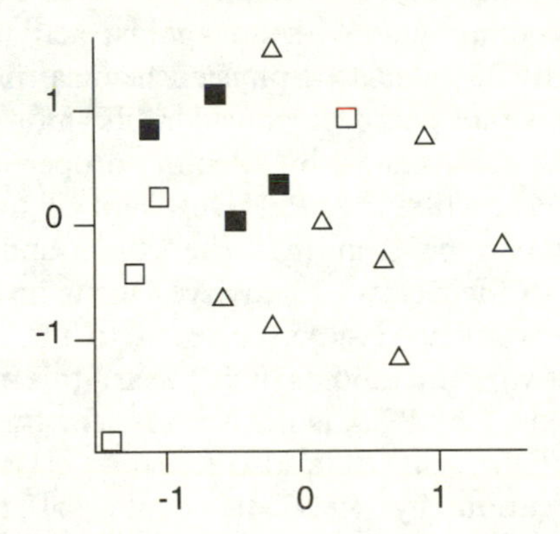

Fig. 9.18. Non-linear map of the same set of compounds as shown in Fig. 9.17 (from Selwood *et al.* 1990, copyright (1990) American Chemical Society).

this case has made a better job of grouping compounds together which have intermediate activity. Can we say that one of these two plots is best? The answer to that depends on the use that is to be made of the display, in other words what questions are we asking of the data. The ReNDeR plot gives a very clear separation between actives and inactives whereas the non-linear map groups most of the intermediates together. An encouraging thing is that the plots produced by the two different display methods are giving similar information about the data. Of course, we might expect that this non-linear display technique would give similar results to another non-linear method such as non-linear mapping. How do ReNDeR plots compare with linear displays such as those produced by principal component analysis (PCA)? Figure 4.8 shows a principal components scores plot for a set of analogues of γ-aminobutyric acid (GABA). These compounds were tested for agonist activity at the central nervous system

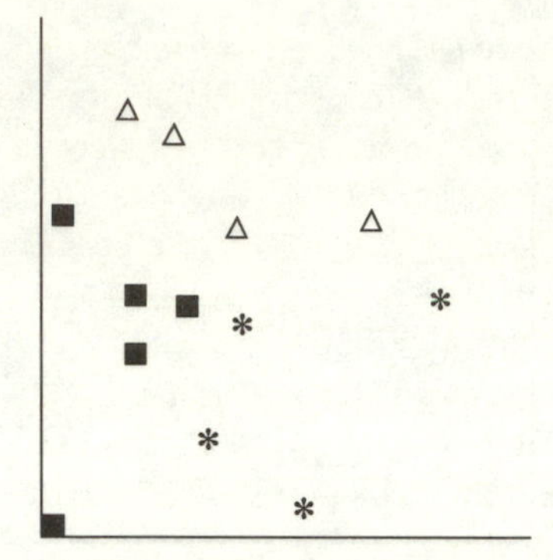

Fig. 9.19. ReNDeR plot of GABA analogues (■ potent agonist, Δ weak agonist, * no agonist activity) described by 33 properties (from Livingstone *et al.* 1991, copyright 1991 by Elsevier Science Inc.).

GABA receptor, and the PC plot roughly separates them into potent and weak agonists and compounds with no agonist activity. The compounds were characterized by 33 calculated physicochemical properties and it was found that the scores plot could be considerably improved, in terms of its ability to classify the compounds, by selecting properties and recomputing the PCA (Hudson *et al.* 1989). A ReNDeR plot of this data is shown in Fig. 9.19 where it can be seen that the compounds are quite clearly grouped according to their class of activity. This is an interesting result in that this technique is giving a superior result to the PCA display of the data. Since the network method is non-linear, this may show that the linear structure imposed by PCA is not suitable for this data set, although a non-linear map of the same data also failed to classify the compounds. The better classification by ReNDeR could, of course, be entirely fortuitous and it will be necessary to examine many other data sets to establish the utility of this technique.

9.3.2 Data analysis using ANN

A neural network may be trained to reproduce any given target from a set of input values, provided it has a sufficient (see later) number of neurons and layers. Where the dependent variable or property to be predicted is classified, e.g., active/intermediate/inactive, the network can be set up with a neuron in the output layer corresponding to each of the classes. Training is carried out until only one neuron (the correct one!) is activated for each of the examples in the training set. This is equivalent to performing discriminant analysis (see Section 7.2.1), physicochemical descriptors are used as input to the network and, once trained, the network connection weights *might* be equated to the coefficients of the parameters in a discriminant function. Unfortunately, the connection weights cannot be identified quite so easily as this since there will be a connection from every input neuron (parameter) to each of the neurons in the hidden layer. After training it may be found that most of these connection weights are around zero and one dominant weight may be found for a particular input parameter. However, it is just as likely that many of the connection weights will have 'significant' values and it will not be possible to extract the contributions made by individual variables.

An example of a data set with a classified response which has been analysed using a neural network is shown in Table 9.7 (Manallack and Livingstone 1992). This data set was chosen because the best discriminant function which could be generated from the sum of the parameters π, MR, and π^2 for the two substituents was only able to classify correctly 22 of the 27 compounds (Prakash and Hodnett 1978). The network architecture used for this analysis consisted of two input units, one for $\Sigma\pi$ and one for

Table 9.7. Structure, physicochemical properties, and activity category of napthoquinones (from Manallack and Livingstone 1992, with permission of Birkhauser)

Compound	R_1	R_2	$\Sigma\pi$	ΣMR	Category[a]
1	Cl	H	0.71	0.70	1
2[b]	OH	H	−0.67	0.38	1
3	OCH_3	H	−0.02	0.89	1
4	$OCOCH_3$	H	−0.64	1.35	1
5[b]	NH_2	H	−1.20	0.64	1
6	NHC_6H_5	H	1.37	3.10	1
7[b]	CH_3	CH_3	1.12	1.12	1
8	CH_3	OCH_3	0.54	1.35	1
9	OH	CH_{3_3}	−0.11	0.84	1
10[b]	Br	Br	1.72	1.78	1
11	Cl	$N(CH_3)_2$	0.89	2.16	1
12	OCH_3	OCH_3	−0.04	1.58	1
13	H	H	0.00	0.20	2
14	CH_3	H	0.56	0.66	2
15[b]	SCH_3	H	0.61	1.48	2
16	Cl	Cl	1.42	1.21	2
17	C_2H_5	H	1.02	1.13	2
18	$COCH_3$	H	−0.55	1.22	1
19[b]	SC_2H_5	H	1.07	1.94	2
20	OH	$CH_2C_6H_5$	1.34	3.28	1
21	OH	$COCH_3$	−1.22	1.4	1
22[b]	CH_3	SCH_3	1.17	1.94	1
23[b]	CH_3	SC_2H_5	1.63	2.4	1
24[b]	OH	Br	0.19	1.17	2
25	Cl	$NHCH_3$	0.24	1.63	1
26	OH	Cl	0.40	0.88	1
27	OH	NH_2	−1.87	0.82	2

[a] Category 1=inactive; 2=active.

[b] Compounds used for testing purposes in the second part of this analysis. Test compounds were chosen at random and the test set possesses approximately the same ratio of inactive to active compounds as in the original data set.

ΣMR, a hidden layer and two output units, so that one unit could be activated (take a positive value) for active compounds and the other unit for inactives. Network training was carried out using networks with different numbers of neurons in the hidden layer so as to assess the performance of the networks. It had already been shown that, given sufficient connections, ANN were able to make apparently successful prediction using random numbers (Manallack and Livingstone 1992). This behaviour of networks is dependent on the number of network connections, the greater the number of connections the more easily (or more completely) a network will train. In fact it was pointed out by Andrea and Kalayeh (1991) that the important quantity is not the overall number of connections but the ratio of the number of data points (samples) to connections which they characterized by the parameter, ρ.

$$\rho = \frac{\text{number of data points}}{\text{number of connections}} \tag{9.4}$$

Table 9.8 shows the effect of adding extra hidden layer neurons to these networks; for a ρ value of 1.00 where there are as many connections as samples the network is able to classify all but two of the compounds successfully. A better test of the suitability of ANN to perform discriminant analysis is to split the data set into separate test and training sets. This was done for this data to give a training set of 18 compounds and a test set of nine as indicated in Table 9.7. Results of training and test set performance are shown in Table 9.9 where it can be seen that training set predictions are good even for the highest value of ρ and at the lower ρ values all compounds are predicted correctly. Prediction performance for the test set, on the other hand, was uniformly bad and if anything got even worse at the lowest values of ρ. This appears to be one of the features of using neural networks to carry out statistical tasks such as discriminant analysis and regression—they generally perform well in prediction of the training set data (i.e., fitting) but poorly when confronted by test set data (i.e., prediction). A recent example (Goodacre *et al.* 1993) of the use of

Table 9.8. Summary of network performance using 27 training compounds (from Manallack and Livingstone 1992, with permission of Birkhauser)

Network architecture	Connections	ρ	Total RMS error	Summary (incorrect compounds)
2, 1, 2	7	3.86	0.4175	1, 7, 15, 17, 19, 24, 27
2, 2, 2	12	2.25	0.3599	1, 7, 8, 27
2, 3, 2	17	1.59	0.2778	1, 7, 8
2, 4, 2	22	1.23	0.2750	1, 7, 8
2, 5, 2	27	1.00	0.2275	17, 24
2, 6, 2	32	0.84	0.1926	8
2, 7, 2	37	0.73	0.0309	All correct

Table 9.9. Summary of network performance and prediction using 18 training compounds (from Manallack and Livingstone 1992, with permission of Birkhauser)

Network architecture	Con*	ρ	Training summary		Prediction summary
			Total RMS error	(incorrect compounds)	(incorrect compounds)
2, 1, 2	7	2.57	0.3166	1, 27	7, 10, 15, 19, 24
2, 2, 2	12	1.50	0.2059	1	2, 5, 7, 10, 15, 19, 24
2, 3, 2	17	1.06	0.3099	1, 27	7, 10, 15, 19, 24
2, 4, 2	22	0.82	0.2359	27	7, 10, 15, 19, 23, 24
2, 5, 2	27	0.67	0.2357	27	7, 10, 15, 19, 23, 24
2, 6, 2	32	0.56	0.0274	All correct	7, 10, 15, 19, 24
2, 7, 2	37	0.49	0.0367	All correct	2, 5, 7, 10, 15, 19, 24

* The number of connections in the network.

neural networks to classify olive oil samples described by pyrolysis mass spectrometry data, however, shows that ANN can work well in prediction. In this work a training set of extra-virgin olive oils and adulterated oil samples (added peanut, sunflower, corn, soya, or sansa olive oils) were analysed by pyrolysis mass spectrometry to give spectra in the M/Z range of 51–200. Cluster analysis and canonical variates analysis of this data showed that the oil samples were broadly classified on the basis of the cultivar from which the extra-virgin oil was derived; extra-virgin and adulterated samples were not distinguished. A three layer back-propagation network with 150 input neurons (one for each M/Z value), eight hidden neurons, and one output neuron was trained with the training set data and found to predict all of the training samples successfully. This is perhaps not surprising since the training set was very small (24 samples) compared with the number of connections in the network (1,217*). Network performance on an unknown test set (samples were analysed blind) was very good, however, as shown in Table 9.10.

ANN may be used to fit a continuous response variable to a set of physicochemical properties, the network just requires one output unit and the training targets are the values of the response variable (IC_{50}, ED_{50}, etc.) for each compound in the set. Performance of these networks, however, can be deceptively good if care is not taken with the network architecture (Livingstone and Manallack 1993). Figure 9.20 shows the results of network training using random numbers in which four columns of random numbers were used as input data and a column of random numbers was used as the target (dependent variable). Any attempt at

* The network has 150 × 8 weights between input and hidden layer, 8 × 1 between hidden and output, and 8 × 1 plus 1 × 1 for the bias units.

Table 9.10. Network prediction of test set oil samples (from Goodacre *et al.* 1993, with permission of the Society of Chemical Industry)

Codename	Network answer[a]	Virgin or adulterated
Perugia	1	Virgin
Lecce	1	Virgin
Urbino	1	Virgin
Rimini	0	Adulterated
Taormina	0	Adulterated
Napoli	1	Virgin
Milano	1	Virgin
Trieste	1	Virgin
Torino	0	Adulterated
Cagliari	0.8[b]	Virgin
Bolzamo	1	Virgin
Venezia	0	Adulterated
Roma	0	Adulterated
Genova	1	Virgin
Bari	1	Virgin
Pescara	0	Adulterated
Padova	0	Adulterated
Palermo	0	Adulterated
Firenze	1	Virgin
Ancona	1	Virgin
Siena	0	Adulterated
Messina	0	Adulterated
Bologna	0	Adulterated

[a] The network was trained and interrogated five times. The scores given are the average of the five runs (± 0.001), where virgin is coded 1 and adulterated oil is coded 0.
[b] The network indicated that the oil Cagliari was of virgin quality (1) on four of the five trainings.

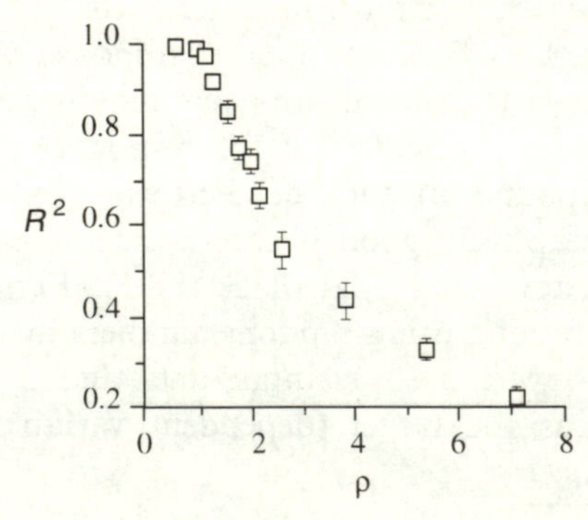

Fig. 9.20. Plot of R^2 versus ρ for regression networks using random numbers (from Livingstone and Manallack 1993, copyright (1993) American Chemical Society).

measuring predictive ability for these networks would be meaningless but the fit can be assessed, as a correlation coefficient (R^2) by comparison of the network output with the target values. As can be seen from the figure, quite high R^2 values are achieved below $\rho = 2$.

9.4 Miscellaneous AI techniques

The expert systems described in Section 9.2 should illustrate some of the principles of the construction and operation of expert systems in chemistry. Given a suitable knowledge base (empirical database) and set of production rules it is possible to predict various chemical properties from structure. Many such systems exist (for example, spectroscopic properties, solubility, heat of formation, etc.), although they are not always called 'expert systems'. The Rekker system for log P (Nys and Rekker 1973) has been coded into a computer-based expert system called PrologP (see Software appendix). This system operates in a very similar way to the CLOGP calculation routine, taking a graphical input of structure, dissecting this into fragments, and then applying the rules of the Rekker scheme to calculate log P. The same company has also produced a pK_a prediction expert system (pKalc) which uses a Hammett equation (for aromatic systems) or a Taft equation (for aliphatics) as a basis for the calculation. Once again, the program takes graphical input of structure which is dissected into fragments. The ionizable groups are perceived and the appropriate equation selected for the prediction of the dissociation constant of each group. The rest of the molecule is treated as fragments or substituents which will modify the pK_a values and fragment constants, equivalent to σ values, are looked-up in a database and applied to the prediction equations. An example of an output screen from this program is shown in Fig. 9.21; the program has the facility to sketch in molecules, as shown for cyclizine, store compounds in a database, and predict pK_a values, as shown for ampicillin. Ionization, of course, affects partition coefficients since it is generally the un-ionized species which partitions into an organic phase.* The PrologP and pKalc programs have been combined to create a distribution coefficient prediction system (PrologD) where log D represents the partitioning of all species at a given pH.

Before leaving expert systems, it is worth considering two problems involved in the prediction of toxicity, namely metabolism and distribution. These two problems are related in that metabolizing enzymes are variously distributed around the body. Thus, the distribution characteristics of a particular compound or its metabolites may dictate which elimination systems they will encounter. Similarly, as metabolism proceeds, the dis-

* Ionized species can dissociate into a 'wet' organic phase singly and as uncharged ion pairs.

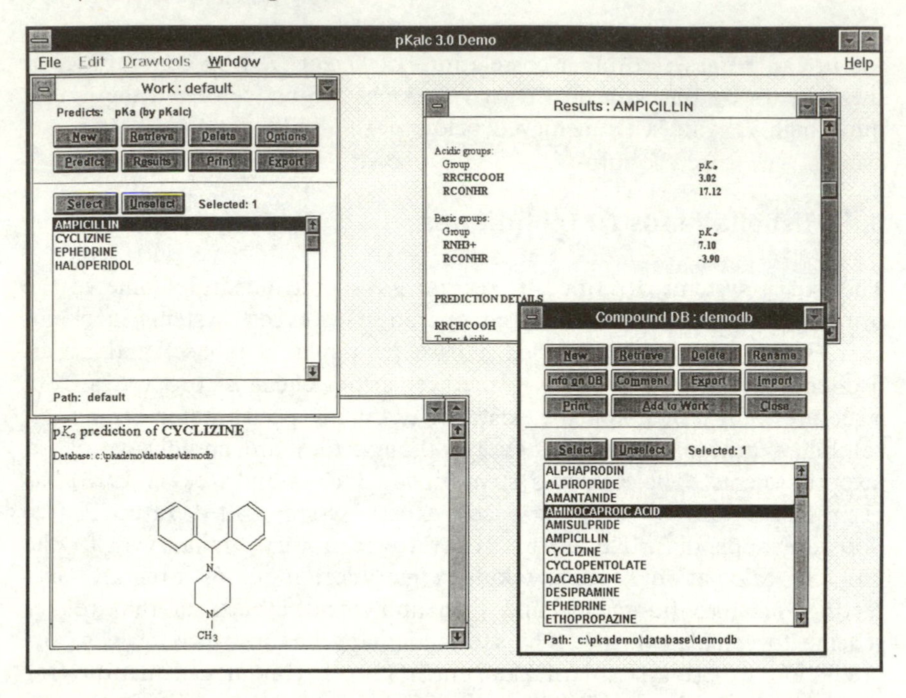

Fig. 9.21. Computer screen from the program pKalc (with permission of Compudrug Chemistry Ltd.).

tribution properties of the metabolites may encourage them to migrate into different tissues. Prediction of toxicity necessarily involves the identification of at least the major products of metabolism and of course the situation is further complicated since different species will be eliminated at different rates. The DEREK system makes some attempt to account for metabolism by incorporating some well-known predictive rules. Other systems deal with specific metabolizing enzymes, for example, the COMPACT program deals with cytochrome P_{450} (Lewis *et al.* 1994), while at least one program (METABOLEXPERT, see Software appendix) attempts to combine metabolic pathways with a simulation of pharmacokinetic behaviour. While these systems have not yet reached the reliability of log *P* prediction programs, an inherently simpler problem, it seems inevitable that they will improve as the body of data is increased.

Rule induction is an artificial intelligence method that has been applied to the analysis of a number of chemical data sets. As the name implies, rule induction aims to extract rules from a set of data (descriptor variables) so as to classify the samples (compounds, objects) into two or more categories. The input to a rule induction algorithm is a number of test cases, a test set, and the output is a tree-structured series of rules, also known as a class probability tree. A popular rule induction algorithm is known as ID3 (Iterative Dichotomizer three) (Quinlan 1986) and its operation in terms of information can be described as follows. If a test set contains p samples of class P and n samples of class N, a sample will belong to class P with probability $p/(p+n)$ and to class N with a probability $n/(p+n)$. The information in a decision tree is given by

$$I(p, n) = -p/(p+n)\log_2 p/(p+n) - n/(p+n)\log_2 n/(p+n) \qquad (9.5)$$

If a particular feature (property, descriptor), F, in the data set, with values $(F_i, F_i + 1 \ldots)$ is used to form the first rule of the decision tree, also known as the 'root' of the tree, it will partition the test set, C, into C_i, C_{i+1}, and so on, subsets. Each subset, C_i, contains those samples which have value F_i of the chosen feature F. If C_i contains p_i samples of class P and n_i samples of class N, the expected information for the subtree C_i is $I(p_i, n_i)$. The expected information required for the tree with feature F as a root is obtained as the weighted average

$$E(F) = \sum_{i=1}^{v} (p_i n_i)/(p+n) I(p_i, n_i) \qquad (9.6)$$

The information gain on branching on feature F is given by eqn (9.7).

$$gain(F) = I(p, n) - E(F) \qquad (9.7)$$

The ID3 procedure examines all the features in the data set and chooses the one that maximizes the gain, this process being repeated until some pre-set number of features are identified or a particular level of reliability is achieved. One problem with this procedure is that 'bushy' trees can be produced, that is to say decision trees which have so many rules that there is a rule for every one or two samples: the ID3 algorithm can be modified, using a significance test, to reject rules that are irrelevant (A-Razzak and Glen 1992).

Examples of the application of the ID3 algorithm to four sets of data involving biologically active compounds have been reported by A-Razzak and Glen (1992). One of these consisted of an expanded version (17 compounds) of the set of 13 γ-aminobutyric acid analogues (GABA) already shown in Figs 4.8 and 9.19. This particular set of compounds was described by seven computed physicochemical properties which did a very reasonable job of separating activity categories, as may be seen from the non-linear map shown in Fig. 9.22. The ID3 algorithm was run on a larger set of 24 computed properties to give the decision tree shown in Fig. 9.23. Interpretation of this tree is fairly self-evident; the data set is split into two above and below a surface area value of 162.15, for example. This is one of the attractions of this form of 'machine learning', the decision rules may be readily understood and should be easy to apply when attempting to design new molecules. Two of the three properties used to provide these

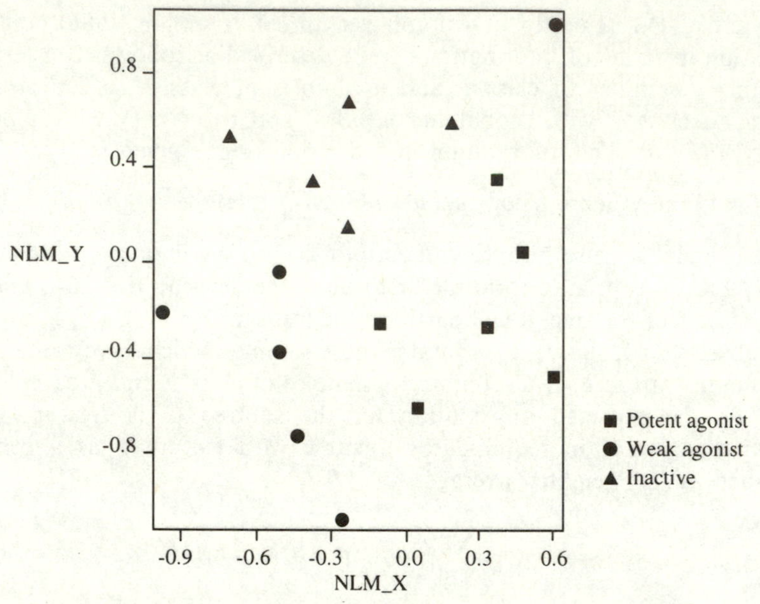

Fig. 9.22. Non-linear map of 17 GABA analogues described by seven physicochemical properties (from A-Razzak and Glen 1992, copyright (1992) Wellcome Foundation and Infolink Decision Services).

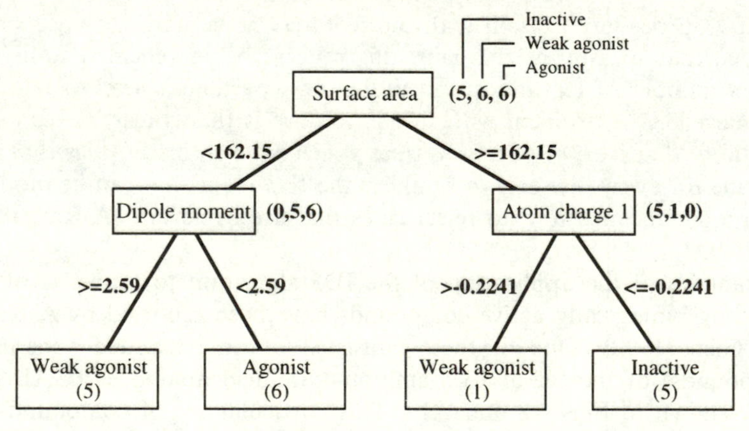

Fig. 9.23. Decision tree from the ID3 algorithm run on the GABA analogues shown in Fig. 9.22 (after A-Razzak and Glen 1992, copyright (1992) Wellcome Foundation and Infolink Decision Services).

decision rules were included in the set of seven parameters used to produce the non-linear map. However, if one is interested in examining the effect of particular variables, perhaps because they have proved important in the prediction of another activity, the ID3 algorithm can be forced to create decision rules for user-selected features. In this example, the samples fell naturally into three classes; where the dependent variable is continuous, the ID3 algorithm can be used by classifying compounds according to a range of the response variable.

Another example of the use of the ID3 algorithm is given in a report which compares 'rule-building expert systems' with pattern recognition in the classification of analytical data (Derde *et al.* 1987). Figure 9.24 shows a principal components plot of 100 olive oil samples characterized by their content of eight fatty acids. The samples are clearly separated into two geographical areas, Eastern and Western Liguria. An implementation of the ID3 algorithm called EX-TRAN (see Software appendix) was able to characterize these samples on the basis of their content of linolenic acid (first rule or root of the tree), oleic acid, linoleic acid, and palmitic acid. A comparison of the performance of EX-TRAN with *k*-nearest-neighbours (one and five neighbours) and linear discriminant analysis (LDA) is shown in Table 9.11, where it can be seen that the results are slightly worse than the pattern recognition techniques (except KNN with raw data).

The final artificial intelligence method for the analysis of chemical data that will be discussed in this section might also be called 'rule-building expert systems'. The widespread use of molecular modelling packages in drug design has led to the creation of 'pharmacophore' or 'biophore' recognizing systems. A pharmacophore is defined as that pattern of atoms (or perhaps properties) which is required to exist in a molecule in order for it to exert some particular biological effect. It is generally accepted that the pharmacophore is 'recognized' by the biological activity site and presumably some, if not all, parts of the pharmacophore are involved in the compound binding to the site. A biophore has a less restrictive definition

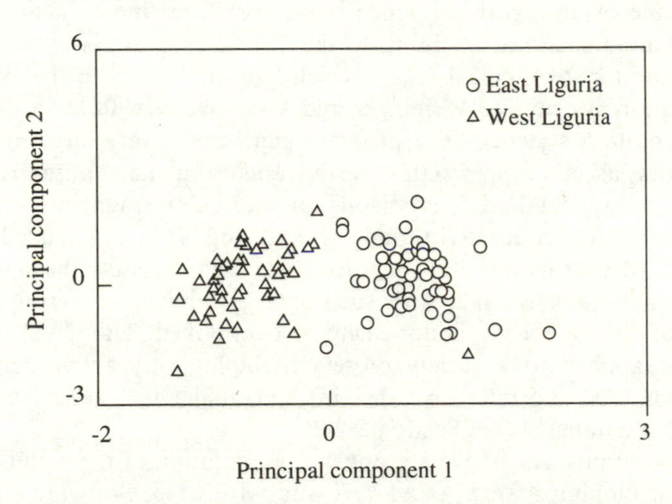

Fig. 9.24. Principal components scores plot for olive oil samples characterized by their content of eight fatty acids (from Derde *et al.* 1987, copyright (1987) American Chemical Society).

Table 9.11. Comparison of validation results: number of wrong prediction results (not counting object 50 of Eastern Liguria) (from Derde *et al.* 1987, copyright (1987) American Chemical Society)

	Eastern Liguria	Western Liguria
EX-TRAN	2	2
1NN		
Raw data	1	5
Autoscaled	1	2
Range-scaled	0	2
5NN		
Raw data	0	5
Autoscaled	1	1
Range-scaled	1	0
LDA		
Stepwise	0	0
All variables	0	1

in that a biophore is some pattern of atoms and/or properties which occurs in some of the active or inactive molecules. The concept of a biophore for inactivity is an interesting one; presumably this relates to a pattern of atoms and/or properties which are responsible for confusing recognition at the active site, or perhaps preventing binding by some repulsive interactions.

The CASE (Computer Assisted Structure Evaluation) program, now elaborated to an enhanced form called MULTICASE (Klopman 1992), is an example of an algorithm which seeks biophores in the active and inactive compounds in a set. Input to the CASE program is by means of a line notation system, called KLN, which has similarities to the Wiswesser line notation system (see Weininger and Weininger, 1990, for a discussion of line notation systems). The program generates a very large number of descriptors, as one report states 'easily ranging in the thousands for 50–100 compound databases', consisting of molecular fragments of various sizes; molecular connectivity indices and log *P* are included in the MULTICASE program. Statistical tests are used to assess the significance of the biophores although, with such a large number of descriptors, the danger of chance effects cannot easily be overlooked. The CASE program has been applied to a variety of sets of biologically active compounds (Klopman 1992 and references therein) and, under the name CASETOX, to toxicity databases (see Section 9.2.2).

Two programs which give a graphical presentation of the results of their search for biophores are CATALYST and APEX (see Software appendix). The APEX program will operate on both two- and three-dimensional chemical structures and in addition to using structural information will

Table 9.12. Atomic indexes calculated with MOPAC used by the biophore recognition program APEX (with permission of Biosym Technologies Ltd.)

Index	Description
ACC–01	Electron-acceptor reactivity of atoms
DON–01	Mean electronic donor reactivity of lone pair
CHARGE	Point atomic charge in atomic units
HOMO	Squares of LCAO coefficients of highest occupied MO
LUMO	Squares of LCAO coefficients of lowest unoccupied MO
P1–POPUL	π-electron density on atoms

also take account of MOPAC calculations, as shown in Table 9.12. The APEX system makes use of two statistics to assess the significance of biophores, probability, $P(A_k/f_r)$, and reliability, P_{rk}, as shown in eqn (9.8) and (9.9).

$$P(A_k/f_r) = \frac{M_{rk}+1}{M_r+2} \tag{9.8}$$

$P(A_k/f_r)$ is the probability that a novel compound possessing a biophore will belong to activity class A_k where M_r is the total number of compounds with the structural pattern f_r and M_{rk} is the number of compounds in activity class A_k which contains that pattern.

$$P_{rk} = \sum_{i=0}^{m_{rk}-1} \binom{m_r}{i} P(A_k)^i \left(1 - P(A_k)\right)^{m_r-1} \tag{9.9}$$

where $P(A_k)$ is the prior probability of the A_k class.

An example of one of the output screens from APEX, from the analysis of a set of mutagens, is shown in Fig. 9.25. Biophores found to be associated with the active set, shown in the box at centre top, are listed in the top right-hand corner of the screen. One of these biophores has been selected for display in the bottom window, and three members of the active set (indicated by highlighting in the active list) are shown superimposed on the biophore. Various information about the biophores, including an assessment of their 'significance' by cross-validation, is easily extracted from the program.

9.5 Summary

Sections 9.2.1 to 9.2.3 described a number of important chemical expert systems which are regularly used in the research and development of pharmaceuticals and agrochemicals. Hopefully, readers with other research

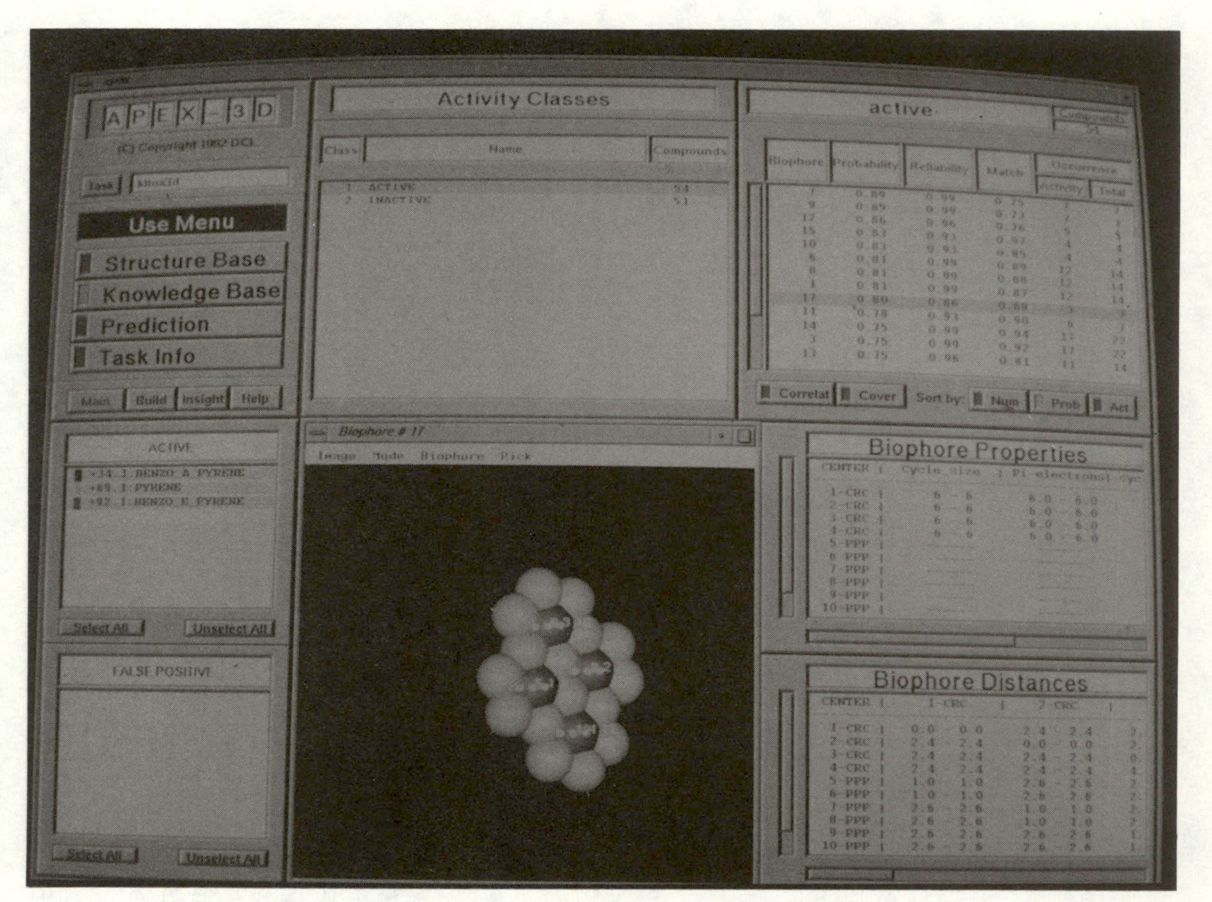

Fig. 9.25. Computer screen from the pharmacophore-recognition program APEX (with permission of Biosym Technologies Ltd.).

or commercial interests may see an application of these and similar systems to their own work, or may even be prompted to develop new systems tailored to their own needs. Section 9.3 discussed neural networks which are a fascinating new development in artificial intelligence techniques and which appear to offer a novel method for the display of multi-dimensional data. Their performance in traditional statistical tasks such as regression and discriminant analysis has so far been disappointing but it remains to be seen how other network architectures may fare in these applications. Finally, rule induction and the pharmacophore/biophore recognition systems appear to show promise in data analysis and should be useful complements to the wide range of more or less well-understood methods available today.

References

Andrea, T. A. and Kalayeh, H. (1991). *Journal of Medicinal Chemistry*, **34**, 2824–36.

A-Razzak, M. and Glen, R. C. (1992). *Journal of Computer-aided Molecular Design*, **6**, 349–83.

Ashby, J. (1985). *Environmental Mutagenesis*, 7, 919–21.

Ashby, J. and Tennant, R. W. (1991). *Mutation Research*, **257**, 229–306.

Ayscough, P. B., Chinnick, S. J., Dybowski, R., and Edwards, P. (1987). *Chemistry and Industry*, **Aug.**, 515–20.

Burns, J. A. and Whitesides, G. M. (1993). *Chemical Reviews*, **93**, 2583–2601.

Cartwright, H. M. (1993). *Applications of artificial intelligence in chemistry*. Oxford University Press.

Corey, E. J. (1991). *Angewandte Chemie*—International edition in English, **30**, 455–65

Corey, E. J., Long, A. K., and Rubenstein, S. D. (1985). *Science*, **228**, 408–18.

Derde, M.-P., Buydens, L., Guns, C., Massart, D. L., and Hopke P. K. (1987). *Analytical Chemistry*, **59**, 1868–71.

Enslein, K., Lander, T. R., Tomb, M. E., and Craig, P. N. (1989). *Toxicology and Industrial Health*, **5**, 265–387.

Enslein, K., Blake, B. W., and Borgstedt, H. H. (1990). *Mutagenesis*, **5**, 305–6.

Goodacre, R., Kell, D. B., and Bianchi, G. (1993). *Journal of the Science of Food and Agriculture*, **63**, 297–307.

Hansch, C. and Leo, A. J. (1979). *Substituent constants for correlation analysis in chemistry and biology*, pp. 18–43. Wiley, New York.

Hileman, B. (1993). *Chemical and Engineering News*, 21 June, 35–7.

Hudson, B., Livingstone, D. J., and Rahr, E. (1989). *Journal of Computer-aided Molecular Design*, **3**, 55–65.

Jakus, V. (1992). *Collection of Czechoslovak Chemical Communications*, **57**, 2413–51.

Judson, P. N. (1992). *Pesticide Science*, **36**, 155–60.

Klopman, G. (1985). *Environmental Health Perspectives*, **61**, 269–74.

Klopman, G. (1992). *Quantitative Structure–Activity Relationships*, **11**, 176–84.

Langowski, J. (1993) *Pharmaceutical Manufacturing International*, 77–80.

Lewis, D. F. V., Moereels, H., Lake, B. G., Ioannides, C., and Parke, D. V. (1994). *Drug Metabolism Reviews* **26**, 261–85.

Livingstone, D. J., Hesketh, G., and Clayworth, D. (1991). *Journal of Molecular Graphics*, **9**, 115–8.

Livingstone, D. J. and Manallack, D. T. (1993). *Journal of Medicinal Chemistry*, **36,** 1295–7.

Livingstone, D. J. and Salt, D. W. (1995). In *Molecular similarity in drug design*, (ed. P. Dean). Chapman & Hall, pp. 187–214.

Manallack, D. T. and Livingstone, D. J. (1992). *Medicinal Chemistry Research*, **2**, 181–90.

Manallack, D. T. and Livingstone, D. J. (1994). In *Advanced computer-assisted techniques in drug discovery*, (ed. H. Van de Waterbeemd), Vol 3 of *Methods and Principles in Medicinal Chemistry* (ed. R. Mannhold, P. Krogsgaard-Larsen and H.Timmerman) pp 293–318. VCH, Weinheim.

Mayer, J. M., Van de Waterbeemd, H., and Testa, B. (1982). *European Journal of Medicinal Chemistry*, **17**, 17–25.

Metivier, P., Gushurst, A. J., and Jorgensen, W. L. (1987). *Journal of Organic Chemistry*, **52**, 3724–38.

Nys, G. C. and Rekker, R. F. (1973). *European Journal of Medicinal Chemistry*, **8**, 521–35.

Quinlan, J. R. (1986). *Machine Learning*, **1**, 81–106.

Prakash, G. and Hodnett, E. M. (1978). *Journal of Medicinal Chemistry*, **21**, 369–73.

Rekker, R. F. and de Kort H. B. (1979). *European Journal of Medicinal Chemistry*, **14**, 479–88.

Salt, D. W., Yildiz, N., Livingstone, D. J., and Tinsley, C. J. (1992). *Pesticide Science*, **36**, 161–70.

Sanderson, D. M. and Earnshaw, C. G. (1991). *Human and Experimental Toxicology*, **10**, 261–73.

Selwood, D. L., Livingstone, D. J., Comley, J. C. W., O' Dowd, A. B., Hudson, A. T., Jackson, P., *et al.* (1990). *Journal of Medicinal Chemistry*, **33**, 136–142.

Tennant, R. W. and Ashby, J. (1991). *Mutation Research*, **257**, 209–27.

Weininger, D. and Weininger, J. L. (1990). In *Quantitative drug design*, (ed. C. A. Ramsden), Vol. 4 of *Comprehensive medicinal chemistry. The rational design, mechanistic study and therapeutic application of chemical compounds* (ed. C. Hansch, P. G. Sammes, and J. B. Taylor), pp. 59–82. Pergamon Press, Oxford.

Zupan, J. and Gasteiger, J. (1991). *Analytica Chimica Acta*, **248,** 1–30.

Zupan, J. and Gasteiger, J. (1993). *Neural networks for chemists*. VCH, Cambridge.

Software appendix

Analysis of any kind of data these days almost invariably involves the use of computers and there is a wide variety of software available running on a range of hardware platforms, from personal computers to mainframes. In addition, for chemical problems, there are specialist packages for the calculation of physicochemical properties, the construction of chemical databases, and the generation of quantitative structure–activity relationships. This appendix is not meant to be a comprehensive list of all the computer programs that might be used in the analysis of chemical data, but does list some of the more commonly used packages and, in particular, all of the programs that have been mentioned in this book. The programs are listed in sections according to their (main) functionality; suppliers and hardware platforms are listed along with a brief description of what each program does. Where the listing indicates a UNIX workstation, the software may run on many different types of machine, or just one; details are available from the supplier.

General-purpose statistics

Most general-purpose statistics packages now include some graph-plotting facilities, e.g., scatter plots, bar charts, pie charts, etc., although this may have to be purchased as an additional module. The following programs all perform the majority of common statistical tasks so no description is given unless they have special or limited functionality

---

ARTHUR	Most computers*	Infometrix Inc., Denny Building, 2200, Sixth Avenue, Suite 833, Seattle, WA 98121, USA

Limited graphics

* Available as FORTRAN source code so potentially may run on any computer with a suitable FORTRAN compiler.

BMDP	IBM mainframe, PC	BMDP, Health Science Computing Facility, AV-111, CHS, University of California, Los Angeles, CA 90024, USA
GENSTAT	VAX, IBM mainframe, PC, UNIX workstation	Numerical Algorithms Group, Mayfield House, 256, Banbury Road, Oxford OX2 7DE, UK
RS/1	VAX, PC	BBN Software Products Corporation, 10, Fawcett Street, Cambridge, MA 02238, USA
Good data-handling capability; limited statistics functionality compared with other packages		
SAS	VAX, PC, UNIX workstation	SAS Institute Inc., SAS Campus Drive, Cary, NC 27513, USA
SIMCA	PC, VAX, UNIX workstation	UMETRI AB, Box 1456, S-90124, Umea, Sweden
PLS and PCA with graphics		
SPSS	IBM mainframe, UNIX workstation, PC	SPSS Inc., Suite 3300, 440, N. Michigan Avenue, Chicago, IL 60611, USA

STATGRAPHICS	PC	STSC Inc., 2115, East Jefferson Street, Rockville, MD 20852, USA
SYSTAT	PC	SYSTAT Inc., 1800, Sherman Avenue, Evanston, IL 60601, USA

Chemical database systems (including reaction databases)

CHEM DBS	VAX, UNIX workstation	Chemical Design Ltd., Roundway House, Cromwell Park, Chipping Norton, Oxfordshire OX7 55R, UK

A module of the CHEM-X modelling system (see modelling section). Storage and retrieval of two- and three-dimensional structures with substructure-search capability. Available databases include: Chapman & Hall Dictionary of Drugs (15,000 compounds), Chapman & Hall Dictionary of Fine Chemicals (120,000 small organics), Chapman & Hall Dictionary of Natural Products (54,000), Derwent Standard Drug File (31,000 biologically active compounds), ChemReact (370,000 reaction types) and others.

MACCS **MACCS 3D**	VAX (optional PC interface)	MDL Information Systems Inc., 14600, Catalina Street, San Leandro, CA 94577, USA

Storage and retrieval of two- and three-dimensional chemical structures with substructure search capability. Available chemical databases include: ACD (available chemicals directory, > 71,000 compounds), Derwent Standard Drug File (37,800 compounds), Drug Data Report (> 20,000 compounds) and a pK_a file (10,700 literature values).

REACCS	VAX	MDL Information Systems Inc., 14600, Catalina Street, San Leandro, CA 94577, USA

A reaction database-searching program. Available databases include: Theilheimer (46,800 reactions), Journal of Synthetic Methods (35,800 reactions), Chiral Synthesis (11,300 reactions), Comprehensive Heterocyclic Chemistry (38,000 reactions) and others.

THOR **(and Merlin)**	VAX* and UNIX workstation	Daylight Chemical Information Systems Inc., 18500, Von Karman Avenue, Suite 450, Irvine, CA 92715, USA

Storage and retrieval of two- and three-dimensional chemical structures with substructure search capability. Available databases include SPRESI (2.2 million compounds), Derwent Drug Index (42,000 drugs), EPA Toxic Substance Control Act ($>$100,000 compounds), and the Pomona College database (26,000 compounds, 9,000 pK_a values, 50,000 P values).

*No longer supported by Daylight.

UNITY	VAX, UNIX workstation	TRIPOS, 1699, S. Hanley Road, Suite 303, St Louis, MO 63144, USA

A module of the SYBYL molecular modelling package (see modelling section). Storage and retrieval of two- and three-dimensional structures, substructure-search, and conformationally flexible search capability. Current databases include the Chapman and Hall dictionaries (Drugs, Fine Chemicals, Natural Products, and Organometallics), Current Drugs, Maybridge catalogue, Cambridge database (crystallographic), SPRESI, and National Cancer Institute (100,000 compounds which have been tested for carcinogenicity).

Chemical property prediction systems

CLOGP, CMR	VAX, UNIX workstation	Daylight Chemical Information Systems Inc., 18500, Von Karman Avenue, Suite 450, Irvine, CA 92715, USA

The CLOGP and CMR modules of the Daylight system (see entry under database software) calculate log *P* by the Hansch and Leo fragment system, and molar refractivity by an atom contribution scheme.

PKALC	PC	Compudrug Chemistry Ltd., H-1136 Budapest, Hollan Erno Utca 5, Hungary e-mail: MKTG@CDK-CGX.HU

pK_a calculation software using Hammett and Taft equations; graphical structure input.

PROLOGP, PROLOGD	PC	Compudrug Chemistry Ltd., H-1136 Budapest, Hollan Erno Utca 5, Hungary e-mail: MKTG@CDK-CGX.HU

Log P (Rekker method) and log D (PROLOGP and pKalc) calculation software.

Toxicity and metabolism prediction systems

CASE, MULTICASE		DSI, 27621, Chagrin Boulevard, Suite 326, Woodmere, Ohio 44122, USA

CASE and MULTICASE are computer programs which relate biological (or other) properties of molecules to chemical structure by means of (mainly) substructural descriptors. MULTICASE is an upgrade to CASE but is not available for distribution; a special version of it, called TOX, along with toxicity database is available.

COMPACT

D. Lewis,
School of Biological Sciences,
University of Surrey,
Guildford,
Surrey GU2 5XH,
UK

A screening system for toxicity via cytochrome P_{450} mediated pathways. Predictions are made on the basis of 'docking' calculations with a P_{450} enzyme model and use of a QSAR model.

DEREK — VAX (in-house or as an on-line service)

LHASA UK Ltd.,
School of Chemistry,
University of Leeds,
Leeds LS2 9JT,
UK

An expert-system-based toxicity prediction program. Toxicity modules include: skin sensitization, FDA rule base, standard rule base (miscellaneous endpoints, e.g., mutagenicity, carcinogenicity), genotoxicity (under development).

METABOLEXPERT — PC

Compudrug Chemistry Ltd.,
H-1136 Budapest,
Hollan Erno Utca 5,
Hungary
e-mail:
MKTG@CDK-CGX.HU

Expert system for the prediction of the metabolic fate of compounds in humans. Results are displayed as a tree-like diagram of the calculated metabolic pathway.

TOPKAT — VAX, PC

Health Designs, Inc.,
183 East Main Street,
Rochester,
NY 14604,
USA

A toxicity-prediction system based on regression or discriminant analysis models built with a variety of databases. Available toxicity modules include: mutagenesis, teratogenesis, carcinogenicity, rat and mouse oral LD_{50}, Fathead Minnow LC_{50}, and others.

Molecular modelling packages

The following molecular modelling packages all allow the construction of models of small molecules and proteins, the minimization of the energy of structures by molecular mechanics, calculation of electronic structure by quantum mechanics, etc. For some programs, certain functionality has to be obtained as an extra module to the basic package, but most of these programs provide most of the functionality that is required for molecular modelling.

Package	Platform	Supplier
CHEM-X	VAX, PC, UNIX workstation, Macintosh	Chemical Design Ltd., Roundway House, Cromwell Park, Chipping Norton, Oxfordshire OX7 55R, UK
INSIGHT	UNIX workstation	Biosym Technologies Inc., 9685, Scranton Road, San Diego, CA 92121-2777, USA
QUANTA	UNIX workstation	Molecular Simulations Inc., 16, New England Executive Park, Burlington, MA 01803, USA
SYBYL	UNIX workstation, PC (stand-alone and as a display for the system running on a server)	TRIPOS, 1699, S. Hanley Road, Suite 303, St Louis, MO 63144, USA

There are a number of other PC and workstation-based systems available.

QSAR packages

CHEMSTAT
A module of the CHEM-X system (see entry under molecular modelling) which allows the calculation of physicochemical properties and the construction of PLS and regression models, PCA and data display.

COMFA
A module of the SYBYL system (see entry under molecular modelling) which characterizes molecules by the calculation of interaction energies of probes at grid points in three-dimensional space. Analysis by PLS and factor analysis.

PARAGON	PC	University of Portsmouth Enterprise Ltd., Town Mount, Portsmouth, Hampshire, PO1 2QG UK

This package includes databases of substituent constant values and (at the time of writing) contains routines for compound selection (various) and the generation of regression models.

TSAR	UNIX workstation	Oxford Molecular Ltd., Oxford Science Park, Sandford-on-Thames, Oxford OX4 4GA UK

This program contains substituent constant data and also calculates physicochemical properties. Analysis facilities include PCA, PLS, regression (MLR), cluster analysis, NLM.

Pharmacophore (toxicophore) recognition systems

APEX
A module of the INSIGHT system (see entry under moleular modelling) which will construct quantitative models based on pharmacophores identified for particular subsets of compounds.

CATALYST
A module of the QUANTA system (see entry under moleular modelling) which will construct quantitative models based on identified pharmacophores.

DISCO
A module of the SYBYL system (see entry under moleular modelling) which identifies the pharmacophores present in an active set of compounds.

REX	VAX	P. Judson, Heather Lea, Bland Hill, Norwood, Harrogate HG3 1TE UK

A pharmacophore recognition system which identifies molecular features present in an active set (compared with a 'reference' set)

Others

CAMEO	VAX	LHASA UK Ltd., School of Chemistry, University of Leeds, Leeds LS2 9JT, UK

Chemical reaction prediction (forwards direction).

LHASA	VAX	LHASA UK Ltd., School of Chemistry, University of Leeds, Leeds LS2 9JT, UK

Retrosynthetic chemical reaction prediction.

EXTRAN	VAX, PC and UNIX workstations	Infolink Decision Services Ltd., 9–11, Grosvenor Gardens, London SW1 W0BD UK

Rule induction system based on the ID3 algorithm.

Neural Networks Almost every type of computer.

There is a very large variety of neural network programs and systems available for almost every type of computing platform. These range from professional packages costing several thousands of pounds to free, public domain software which can often be obtained (by anonymous FTP or similar) over the internet. Two useful sources of information are the following electronic bulletin boards:

Central Neural System BBS has an electronic bulletin board containing 26 Mbytes of files related to artificial neural networks including simulation packages, demos, source code, tutorials and other text. Most is suited to IBM PC compatible machines but there is some available for Macintosh and Unix machines. CNS BBS can be contacted via Wesley Elsberry, P.O. Box 1187, Richland, WA 99352, USA; email: elsberry@beta.tricity.wsu.edu.

The Neuron Digest bulletin board reports (approx. weekly) on various aspects of neural network activities and can be accessed by contacting the moderator Peter Marvit on the following email address; marvit@cattell.psych.upenn.edu.

The software used to produce the results shown in Chap. 9 was:

BIOPROP. Bioprop is a programmable neural network simulator which uses a command language. Several example scripts are provided with the manual which can be obtained from The Office of Technology and Licensing, University of California, Berkeley, CA 94709, USA.

ReNDeR. The ReNDeR program may be used to produce non-linear displays and principal components plots of data sets (as well as regression and classification). Further details are available from Andy Lewcock, AEA Technology, Applied Neurocomputing Centre, 8.12 Harwell, Didcot, Oxon OX11 0RA, UK

A selection of books which give code (some on disk) and advice on the practical application of neural network models is:

Korn, G. A. (1991) *Neural Network Experiments on Personal Computers and Workstations*, MIT Press, Cambridge, MA.

Caudill, M. and Butler, C. (1991) *Understanding Neural Networks*, Vol I and II, MIT Press, Cambridge, MA.

Eberhart, R. C. and Dobbins, R. W. (1990) *Neural Network PC Tools*, Academic Press, Cambridge, MA: Disk for book from Software Frontiers; Gilbert, A. Z.

McCord, N. M. and Illingworth, W. T. (1990) *A Practical Guide to Neural Nets*, Addison-Wesely, Reading, MA.

Aleksander, I, (1990) *An Introduction to Neural Computing*, Chapman and Hall, London. Software for book from Adhoc Reading Systems; East Brunswick, NJ, USA.

Blum, A. (1992) *Neural Networks Programming in C++*, John-Wiley, New York, NY.

Masters, T. (1993) *Practical Neural Network Recipes in C++*, Academic Press, New York, NY.

Finally, a list of free software and frequently asked questions (FAQ) can be obtained from Lutz Prechelt, University of Karlsruhe, Germany, email; prechelt@ira.uka.de

Index

Where several entries are given, bold type indicates key pages.